Selected Titles in This Series

56 **Leonid I. Korogodski and Yan S. Soibelman,** Algebras of functions on quantum groups: Part I, 1998

55 **J. Scott Carter and Masahico Saito,** Knotted surfaces and their diagrams, 1998

54 **Casper Goffman, Togo Nishiura, and Daniel Waterman,** Homeomorphisms in analysis, 1997

53 **Andreas Kriegl and Peter W. Michor,** The convenient setting of global analysis, 1997

52 **V. A. Kozlov, V. G. Maz'ya, and J. Rossmann,** Elliptic boundary value problems in domains with point singularities, 1997

51 **Jan Malý and William P. Ziemer,** Fine regularity of solutions of elliptic partial differential equations, 1997

50 **Jon Aaronson,** An introduction to infinite ergodic theory, 1997

49 **R. E. Showalter,** Monotone operators in Banach space and nonlinear partial differential equations, 1997

48 **Paul-Jean Cahen and Jean-Luc Chabert,** Integer-valued polynomials, 1997

47 **A. D. Elmendorf, I. Kriz, M. A. Mandell, and J. P. May (with an appendix by M. Cole),** Rings, modules, and algebras in stable homotopy theory, 1997

46 **Stephen Lipscomb,** Symmetric inverse semigroups, 1996

45 **George M. Bergman and Adam O. Hausknecht,** Cogroups and co-rings in categories of associative rings, 1996

44 **J. Amorós, M. Burger, K. Corlette, D. Kotschick, and D. Toledo,** Fundamental groups of compact Kähler manifolds, 1996

43 **James E. Humphreys,** Conjugacy classes in semisimple algebraic groups, 1995

42 **Ralph Freese, Jaroslav Ježek, and J. B. Nation,** Free lattices, 1995

41 **Hal L. Smith,** Monotone dynamical systems: an introduction to the theory of competitive and cooperative systems, 1995

40.3 **Daniel Gorenstein, Richard Lyons, and Ronald Solomon,** The classification of the finite simple groups, number 3, 1998

40.2 **Daniel Gorenstein, Richard Lyons, and Ronald Solomon,** The classification of the finite simple groups, number 2, 1995

40.1 **Daniel Gorenstein, Richard Lyons, and Ronald Solomon,** The classification of the finite simple groups, number 1, 1994

39 **Sigurdur Helgason,** Geometric analysis on symmetric spaces, 1994

38 **Guy David and Stephen Semmes,** Analysis of and on uniformly rectifiable sets, 1993

37 **Leonard Lewin, Editor,** Structural properties of polylogarithms, 1991

36 **John B. Conway,** The theory of subnormal operators, 1991

35 **Shreeram S. Abhyankar,** Algebraic geometry for scientists and engineers, 1990

34 **Victor Isakov,** Inverse source problems, 1990

33 **Vladimir G. Berkovich,** Spectral theory and analytic geometry over non-Archimedean fields, 1990

32 **Howard Jacobowitz,** An introduction to CR structures, 1990

31 **Paul J. Sally, Jr. and David A. Vogan, Jr., Editors,** Representation theory and harmonic analysis on semisimple Lie groups, 1989

30 **Thomas W. Cusick and Mary E. Flahive,** The Markoff and Lagrange spectra, 1989

29 **Alan L. T. Paterson,** Amenability, 1988

28 **Richard Beals, Percy Deift, and Carlos Tomei,** Direct and inverse scattering on the line, 1988

27 **Nathan J. Fine,** Basic hypergeometric series and applications, 1988

26 **Hari Bercovici,** Operator theory and arithmetic in H^∞, 1988

25 **Jack K. Hale,** Asymptotic behavior of dissipative systems, 1988

(See the AMS catalog for earlier titles)

Algebras of Functions on Quantum Groups: Part I

Mathematical Surveys and Monographs

Volume 56

Algebras of Functions on Quantum Groups: Part I

Leonid I. Korogodski
Yan S. Soibelman

American Mathematical Society

Editorial Board

Georgia M. Benkart Tudor Stefan Ratiu, Chair
Howard A. Masur   Michael Renardy

1991 *Mathematics Subject Classification.* Primary 17B37; Secondary 16W30, 81R50.

ABSTRACT. The book is devoted to the study of algebras of functions on quantum groups. The authors' approach to the subject is based on the parallels with symplectic geometry, allowing the reader to use geometric intuition in the theory of quantum groups. The book includes the theory of Poisson–Lie algebras (quasi-classical version of algebras of functions on quantum groups), a description of representations of algebras of functions, and the theory of quantum Weyl groups.

The book can serve as an introduction to the theory of quantum groups. It can be used by graduate students and researchers working in algebra, representation theory, and mathematical physics.

Library of Congress Cataloging-in-Publication Data
Korogodski, Leonid I.
   Algebras of functions on quantum groups / Leonid I. Korogodski, Yan S. Soibelman.
     p. cm. — (Mathematical surveys and monographs, ISSN 0076-5376 ; v. 56)
   Includes bibliographical references.
   ISBN 0-8218-0336-0 (pt. I)
   1. Quantum groups. 2. Function algebras. I. Soibelman, Yan S. II. Title. III. Series: Mathematical surveys and monographs ; no. 56.
QC20.7.G76K67 1998
512′.55—dc21   98-10602
                                                                                   CIP

Copying and reprinting. Individual readers of this publication, and nonprofit libraries acting for them, are permitted to make fair use of the material, such as to copy a chapter for use in teaching or research. Permission is granted to quote brief passages from this publication in reviews, provided the customary acknowledgment of the source is given.

Republication, systematic copying, or multiple reproduction of any material in this publication (including abstracts) is permitted only under license from the American Mathematical Society. Requests for such permission should be addressed to the Assistant to the Publisher, American Mathematical Society, P. O. Box 6248, Providence, Rhode Island 02940-6248. Requests can also be made by e-mail to reprint-permission@ams.org.

© 1998 by the American Mathematical Society. All rights reserved.
The American Mathematical Society retains all rights
except those granted to the United States Government.
Printed in the United States of America.

∞ The paper used in this book is acid-free and falls within the guidelines
established to ensure permanence and durability.
Visit the AMS home page at URL: http://www.ams.org/

Contents

Chapter 0. Introduction — 1

Chapter 1. Poisson Lie Groups — 5
  1. Poisson manifolds — 5
    1.1. Poisson algebras and Poisson manifolds — 5
    1.2. Symplectic leaves in a Poisson manifold — 7
  2. Lie bialgebras and Manin triples — 10
    2.1. Lie bialgebras — 10
    2.2. Co-Poisson Hopf algebras — 12
    2.3. Manin triples — 14
  3. Poisson groups — 17
    3.1. Poisson affine algebraic groups and Poisson Hopf algebras — 17
    3.2. Poisson Lie groups — 18
    3.3. The correspondence between Poisson Lie groups and Lie bialgebras — 20
    3.4. Symplectic leaves in Poisson Lie groups and dressing action — 23
  4. Lie bialgebras and classical r-matrices — 27
    4.1. Coboundary, quasi-triangular and triangular Lie bialgebras — 27
    4.2. The classification of quasi-triangular Lie bialgebras — 30
    4.3. Frobenius Lie algebras and CYBE — 35
  5. Compact Poisson Lie groups — 37
    5.1. Quasi-triangular compact Poisson Lie groups — 37
    5.2. Symplectic leaves and Bruhat decomposition — 38
    5.3. Symplectic leaves in simple complex Poisson Lie groups — 42
  6. Poisson G-manifolds — 43
    6.1. Poisson G-manifolds and moment maps — 43
    6.2. Poisson homogeneous G-manifolds — 49
  7. Historical remarks — 55

Chapter 2. Quantized Universal Enveloping Algebras — 57
  1. Quantization of Lie bialgebras — 57
    1.1. Definition of quantization — 57
    1.2. The quantization of complex simple Lie algebras — 58
    1.3. Existence and uniqueness of quantization — 61
  2. QUE-algebras and R-matrices — 61
    2.1. Types of Hopf algebras — 62
    2.2. Double Hopf algebras — 64
    2.3. The quantum double and the universal quantum R-matrix — 65
    2.4. Twisted version of $U_h\mathfrak{g}$ — 71

3.	Center of quasi-triangular Hopf algebras	72
	3.1. Two central element constructions	72
	3.2. Square of the antipode in the almost-cocommutative case	74
	3.3. The quasi-triangular case	76
4.	Center of $U_h\mathfrak{g}$ and quantum Harish-Chandra homomorphism	78
	4.1. Central elements of $U_h\mathfrak{g}$	79
	4.2. Quantum Harish-Chandra homomorphism	80
5.	Finite-dimensional $U_h\mathfrak{g}$-modules	82
	5.1. Finite-dimensional modules and highest weights	82
	5.2. Central characters and the quantum Harish-Chandra homomorphism	85
6.	Tensor products of $U_h\mathfrak{g}$-modules and tensor categories	86
7.	Fixed quantization parameter	89
	7.1. The complex Hopf algebra $U_q\mathfrak{g}$	90
	7.2. Quasi-R-matrix	90
	7.3. Admissible finite-dimensional $U_q\mathfrak{g}$-modules	92
	7.4. Twisted version of $U_q\mathfrak{g}$	93
8.	Historical remarks	94

Chapter 3. Quantized Algebras of Functions   95
  1. Main definitions   95
    1.1. Hopf $*$-algebras   95
    1.2. Quantized algebra of regular functions   96
  2. Properties of the quantized algebras of functions   97
    2.1. Basic properties   97
    2.2. Triangular decomposition of $\mathbb{C}[G]_q$ 99
    2.3. The involution $*$ in $\mathbb{C}[K]_q$ 100
  3. Examples: $\mathbb{C}[SL_2(\mathbb{C})]_q$ and $\mathbb{C}[SU(2)]_q$ 101
  4. Representation theory of $\mathbb{C}[SU(2)]_q$ 104
    4.1. Unitarizable simple $\mathbb{C}[SL_2(\mathbb{C})]_q$-modules   105
    4.2. Irreducible $*$-representations of $\mathbb{C}[SU(2)]_q$ 107
  5. Representation theory of $\mathbb{C}[K]_q$ 109
    5.1. Highest weight modules and primitive ideals   109
    5.2. Primitive ideals and Schubert cells   111
    5.3. Representations of $\mathbb{C}[K]_q$ and Schubert cells   113
  6. Representations of $\mathbb{C}[K]_q$ and symplectic leaves   116
    6.1. Elementary representations of $\mathbb{C}[K]_q$ 116
    6.2. Tensor product theorem   119
  7. Representation theory of the twisted algebras of functions   122
    7.1. Twisted quantized algebras of functions   122
    7.2. Representations of $\mathbb{C}[K^u]_q$ and quantum tori   123
    7.3. Representations of $\mathbb{C}[K^u]_q$ and symplectic leaves in K^u 127
    7.4. Representations of $\mathbb{C}[K(0,u)]_q$ 128
  8. Representations of formal quantized algebras of functions   130
  9. Historical remarks   131

Chapter 4. Quantum Weyl Group and the Universal Quantum R-Matrix   133
  1. Motivations: Weyl group in the quasi-classical picture   133

2.	Quantum Weyl group: definitions	135
	2.1. The case of $\mathbb{C}[SL_2(\mathbb{C})]_{q_i}$	136
	2.2. General case	138
	2.3. Formal case	139
3.	Quantum root vectors	140
	3.1. Poincaré–Birkhoff–Witt theorem	140
	3.2. Properties of quantum root vectors	141
4.	The universal quantum R-matrix	143
	4.1. The PBW-basis and the universal quantum R-matrix	143
	4.2. The case of fixed q	145
5.	Applications to the representations of the braid groups	145
6.	Historical remarks	146

Bibliography   149

CHAPTER 0

Introduction

0.1. This is the first part of the book devoted to the algebras of functions on quantum groups. It is based in part on the lectures the second author gave in England in 1991, at the European School of Group Theory in Netherlands in 1992, and on his Spring lecture course at Harvard in 1996. The aim of the lectures was to present the second author's results on the representation theory of algebras of functions on compact quantum groups including various applications (quantum Weyl groups, universal quantum R-matrices, q-special functions, etc.).

The book is based on the original approach to quantum groups developed by V. Drinfeld. Although in general the book might preserve the lecture course style, we believe that it also can be used as a systematic introduction to quantum groups via Drinfeld's approach.

0.2. Before describing the content of the book we would like to say a few general words about the topic. We address these words to beginners; a more advanced reader can skip this section.

Quantum groups were introduced in the early 1980s. One of the main purposes was to uncover mathematical structures of the Quantum Inverse Scattering Method developed by the school of L. Faddeev (key words: Yang–Baxter equation, R-matrix, transfer matrix). Soon, it was realized that the new object had nontrivial applications not only to various areas of physics (as, for example, the Conformal Field Theory), but also to numerous branches of mathematics (representation theory, topology, analysis, combinatorics, etc.).

These mathematical applications were mostly based on the quantized universal enveloping algebras (which are certain deformations of the universal enveloping algebras of usual Lie algebras). Algebras of functions on quantum groups are objects dual to those. From the point of view of noncommutative geometry they should be studied first, since in the classical case their spectra contain all the information about groups. This was hard to do for at least two reasons. First, there is no good description of the coordinate ring of, say, a complex simple Lie group. While it can be described in terms of generators and relations, such descriptions differ for different series of simple Lie groups. On the other hand, there is a general description of the universal enveloping algebra of the corresponding Lie algebra (in terms of the Chevalley generators), so that the Peter–Weyl theorem recovers the coordinate ring in terms of matrix elements of its finite-dimensional representations. Second, finite-dimensional representations of most familiar quantized universal enveloping algebras form a braided monoidal category and this fact is in the core of many applications of quantum groups (for example to topology). The corresponding quantized algebras of functions do not have this property.

Despite these difficulties, mathematicians have developed a technique to work with quantized algebras of functions. They have been rewarded by a number of nontrivial applications. Although these applications will be the subject of the second part of the book, we mention two of them here: the discovery of the relation between quantum groups and q-special functions by L. Vaksman and the second author in 1986, and the discovery of the connection between the Zamolodchikov tetrahedra equation and quantum algebras of functions at roots of 1 by M. Kapranov and V. Voevodsky, and by D. Kazhdan and the second author in 1992.

The whole range of topics should include also harmonic analysis on quantum groups, q-difference equations, the quantum orbit method, etc., in particular, the results of the first author on the still mysterious subject of noncompact quantum groups.

0.3. There are several approaches to quantum groups. We do not intend to compare them in detail in this book. Instead, we focus on Drinfeld's approach, which is, in our opinion, underrepresented in the existing textbooks.

One of its strengths is in allowing parallels with the symplectic geometry, which thus becomes a powerful source of intuition in quantum groups. In fact, the essence of that approach is in comparing two pictures: the quantum picture and the so-called "quasi-classical" picture. The main premise is the concept that any result on the quantum level has its quasi-classical analog. Therefore, for any quasi-classical result, one can ask how to quantize it. Although this idea has a long history, it is still fruitful. In quantum groups it is supported by the fact that in many cases the quantization parameter can be fixed. This produces interesting phenomena, such as discretization of the space with the typical length equal to the quantization parameter.

Now we describe the content of the book.

We start with Chapter 1 devoted to Poisson Lie groups, the quasi-classical objects that are to be quantized. This chapter contains the results we have borrowed from different works. Although many of them now have simpler proofs (especially if one uses the notion of Lie bialgebroid and Poisson Lie groupoid developed by A. Weinstein and his collaborators), we have decided to keep the older versions in most cases, waiting for the theory of quantization of the newer objects.

Chapter 2 is mainly based on Drinfeld's results on quantization of Lie bialgebras. Those results are taken from his papers. Since we use these results all the time, we have decided to combine them in a separate chapter. Chapter 3 is based on the results of the second author with few later improvements. Again, we have decided to keep the older proofs, which might be less algebraic than the later versions, but have more transparent relations to symplectic geometry of Poisson Lie groups, tracing the correspondence between the algebra of the quantum picture and the geometry of the underlying quasi-classical one. Chapter 4 is devoted to the first applications of the results of the previous chapters: the quantum theory of the Weyl group and the universal quantum R-matrix. It starts with the quasi-classical motivation due to Drinfeld. Then we use the representation theory of the algebras of functions developed in Chapter 3 to define the quantized Weyl group. This approach gives a conceptual explanation of the braid relations between quantized simple reflections.

Each chapter starts with a short review of its content. At the end of each chapter we give some historical remarks, paying tribute to numerous mathematicians whose contribution to the theory of quantum groups we use in the chapter.

These remarks are obviously incomplete, and we apologize for missing names and references.

The book contains many exercises. Some of them are trivial, others include results from original papers. These exercises form an important part of the book (sometimes we refer to them later) and we recommend that the reader work on most of them.

ACKNOWLEDGEMENTS. We are deeply grateful to Vladimir Drinfeld who in 1985 started to explain to the second author the hitherto young concept of a quantum group. Drinfeld's influence on this book is difficult to overestimate. The first author was under this influence when he joined the team formed by the second author and Leonid Vaksman, his teacher, to whom he owes more than simple words of gratitude.

We express our gratitude to our collaborators, especially Leonid Vaksman and Serge Levendorskii; without joint works with them this book would have never appeared. We thank all our colleagues and students for hundreds of discussions on this exciting subject — the quantum groups.

The work of the first author was supported by NSF grant DMS 9304580.

CHAPTER 1

Poisson Lie Groups

In this chapter we discuss the notion of Poisson Lie group and Lie bialgebra. Manin triples give a tool for constructing examples. After a discussion on dressing transformations we apply this notion to the description of symplectic leaves in Poisson Lie groups. Thus the classical Bruhat decomposition follows from the general fact that any Poisson manifold is a union of its symplectic leaves. A short account of Poisson homogeneous spaces is given at the very end of the chapter.

Throughout the book we assume that **k** is a field of characteristic zero unless stated otherwise.

1. Poisson manifolds

In the present section we recall some basic facts about Poisson manifolds. All the material summarized here is well known. The interested reader can find a more detailed and profound treatment in a vast variety of articles and monographs dealing exclusively with this subject. Some of the sources are mentioned in the references.

1.1. Poisson algebras and Poisson manifolds.

DEFINITION 1.1.1. A commutative associative algebra A over **k** is called a *Poisson algebra* if it is equipped with a **k**-bilinear operation $\{\,,\,\} : A \otimes A \to A$ such that the following conditions are satisfied:

(1) A is a Lie algebra with the bracket $\{\,,\,\}$;
(2) the Leibniz rule is satisfied, i.e. for any $a, b, c \in A$, we have

(1.1.1) $$\{ab, c\} = a\{b, c\} + \{a, c\}b.$$

Also, if these conditions are satisfied, the operation $\{\,,\,\}$ is called *Poisson bracket*.

REMARK 1.1.2. Note that (1.1.1) is equivalent to the fact that $\xi_a = \{a, \cdot\}$ is a derivation of A. ξ_a is called *Hamiltonian derivation*.

DEFINITION 1.1.3. Let A and B be Poisson algebras over **k**. An algebra homomorphism $f : A \to B$ is called *Poisson homomorphism* if

$$f(\{a, b\}) = \{f(a), f(b)\}.$$

The Poisson algebras form a category, with morphisms being Poisson homomorphisms.

REMARK 1.1.4. The category of Poisson algebras with unity can be defined in the same way with the only difference that the morphisms are required to be unital, that is $f(1) = 1$.

DEFINITION 1.1.5. A smooth manifold M is called a *smooth Poisson manifold* if the algebra $A = C^\infty(M)$ of smooth complex-valued functions on M is equipped with a structure of Poisson algebra over $\mathbb{C}$.

DEFINITION 1.1.6. (1) An affine algebraic **k**-variety M is called an *affine algebraic Poisson* **k**-*variety* if the algebra $A = \mathbf{k}[M]$ of regular functions on M is equipped with a structure of Poisson algebra over **k**.

(2) An algebraic **k**-variety M is called an *algebraic Poisson* **k**-*variety* if the sheaf $\mathcal{O}_M$ of regular functions carries a structure of a sheaf of Poisson algebras.

DEFINITION 1.1.7. Let M and N be smooth Poisson manifolds (resp. algebraic Poisson **k**-manifolds). A smooth (resp. algebraic) map $f : M \to N$ is called a *Poisson map* if the induced map $f^* : C^\infty(N) \to C^\infty(M)$ (resp. $f^* : \mathbf{k}[N] \to \mathbf{k}[M]$) is a Poisson homomorphism.

It is clear that smooth Poisson manifolds form a category, as do (affine) algebraic Poisson **k**-varieties. Throughout this section we deal only with smooth Poisson manifolds, so that we would omit the adjective 'smooth'.

DEFINITION 1.1.8. Suppose that M is a Poisson manifold.

(1) Given $\varphi \in C^\infty(M)$, the vector field ξ_φ on M associated to the Hamiltonian derivation $\{\varphi, \cdot\}$ of C^∞ is called *Hamiltonian vector field*.

(2) A submanifold (not necessarily closed) $N \subset M$ is called *Poisson submanifold* if the vector $\xi_\varphi(n)$ is tangent to N for any $n \in N$ and $\varphi \in C^\infty(M)$.

There is an equivalent definition of Poisson manifolds in terms of bivector fields. Recall that an n-vector field is a section of the vector bundle $\bigwedge^n TM$ where TM is the tangent vector bundle over a manifold M. In particular, we just call 2-vector fields *bivector fields*.

Recall also the definition of the Schouten bracket of n-vector fields which generalizes the usual Lie bracket of vector fields. The Schouten bracket of an m-vector field and an n-vector field is an $(m + n - 1)$-vector field which is locally defined as follows:

$$[u_1 \wedge u_2 \wedge \cdots \wedge u_m, v_1 \wedge v_2 \wedge \cdots \wedge v_n]$$
$$= \sum_{i,j} (-1)^{i+j} [u_i, v_j] \wedge u_1 \wedge \cdots \wedge \hat{u}_i \wedge \cdots \wedge u_m \wedge v_1 \wedge \cdots \wedge \hat{v}_j \wedge \cdots \wedge v_n,$$

where $u_1, \ldots, u_m, v_1, \ldots, v_n \in T_m M$, $m \in M$. The usual convention $\hat{a}$ means that a does not occur in the formula at that particular place.

Denote by T^*M the cotangent bundle to M. Given a bivector field π on M, we define a bilinear operation $\{\,,\,\}$ on $C^\infty(M)$ by the formula

(1.1.2) $$\{\varphi, \psi\} = \langle \pi, d\varphi \wedge d\psi \rangle,$$

where $\langle\,,\,\rangle$ is the natural pairing between the sections of the bundles $T^*M \wedge T^*M$ and $TM \wedge TM$.

PROPOSITION 1.1.9. *The bracket* (1.1.2) *defines a Poisson manifold structure on M if and only if*

(1.1.3) $$[\pi, \pi] = 0.$$

PROOF. The skew-commutativity of the bracket (1.1.2) is obvious. The Leibniz rule (1.1.1) is satisfied due to the formula $d(\varphi\psi) = \varphi d\psi + \psi d\varphi$. Finally, it is a straightforward computation that the Jacobi identity is equivalent to (1.1.3). $\square$

Consider the morphism of vector bundles $\check{\pi} : T^*M \to TM$ induced by π. Then for any $\varphi \in C^\infty(M)$, we have
$$\xi_\varphi = \check{\pi}(d\varphi).$$

DEFINITION 1.1.10. A Poisson manifold M is called *symplectic* if the map $\check{\pi} : T^*M \to TM$ is an isomorphism.

REMARK 1.1.11. Let M be a symplectic manifold. Then there exists a closed nondegenerate differential 2-form ω such that
$$\omega(\xi_\varphi, \xi_\psi) = \{\varphi, \psi\}$$
for any $\varphi, \psi \in C^\infty(M)$. The form ω is the image of the bivector field π with respect to the isomorphism between $T^*M \wedge T^*M$ and $TM \wedge TM$ induced by π.

Hence, a symplectic manifold is just a pair (M, ω), where ω is a nondegenerate closed differential 2-form on a smooth manifold M.

EXAMPLE 1.1.12. Consider the manifold $M = \mathbb{R}^{2n}$ equipped with the closed 2-form

(1.1.4) $$\omega = \sum_{i=1}^n dp_i \wedge dq_i,$$

where $p_1, \ldots, p_n, q_1, \ldots, q_n$ are the coordinates in $\mathbb{R}^{2n}$.

By the Darboux theorem, any point of a $2n$-dimensional symplectic manifold has a neighborhood with some coordinates $p_1, \ldots, p_n, q_1, \ldots, q_n$ (called *canonical*) such that the symplectic form ω is given by (1.1.4).

1.2. Symplectic leaves in a Poisson manifold. One of the most fundamental facts in the theory of Poisson manifolds is that for any Poisson manifold M there is a stratification of M by symplectic submanifolds which are called symplectic leaves in M. In other words, symplectic manifolds are simple objects in the category of Poisson manifolds (the morphisms in the category of Poisson manifolds are Poisson maps).

In what follows we assume that M is a smooth Poisson manifold.

DEFINITION 1.2.1. A *Hamiltonian curve* on a smooth Poisson manifold M is a smooth curve $\gamma : [0, 1] \to M$ such that there exists $f \in C^\infty(M)$ with the property that
$$\dot{\gamma}(t) = \xi_f(\gamma(t))$$
for any $t \in (0, 1)$.

DEFINITION 1.2.2. Let M be a smooth Poisson manifold.
(1) We say that two points $x, y \in M$ are *equivalent* if they can be connected by a piecewise Hamiltonian curve.
(2) An equivalence class of points of M is called *symplectic leaf* of M.

PROPOSITION 1.2.3. *Let S be a symplectic leaf of a smooth Poisson manifold M. Then:*
  (1) *S is a Poisson submanifold of M;*
  (2) *S is a symplectic manifold;*
  (3) *M is a union of its symplectic leaves.*

PROOF. Let S be a symplectic leaf. Recall that the bivector field π defines a morphism $\check{\pi}_x : T_x^*M \to T_xM$ for any $x \in M$. The Jacobi identity for the Poisson brackets implies the invariance of $\check{\pi}$ along a symplectic leaf. Therefore, the rank of $\check{\pi}_x$ is constant for any $x \in S$. Let r be the dimension of the image of $\check{\pi}_x$ in T_xM for any $x \in S$. One can always choose a set of functions $f_1, f_2, \ldots, f_r$ on M such that the Hamiltonian vector fields $\xi_{f_1}, \xi_{f_2}, \ldots, \xi_{f_r}$ form a basis of $\check{\pi}_x(T_x^*M)$ for any $x \in S$. This means that $d_x f_1, d_x f_2, \ldots, d_x f_r$ are linearly independent, and the restrictions of the functions $f_1, f_2, \ldots, f_r$ to S define a coordinate system in a neighborhood of $x \in S$ in S. We leave it as an easy exercise to complete the proof of (1) by showing that the transition functions between different local coordinate systems are smooth.

Given any functions φ and ψ on S, extend them to functions $\bar{\varphi}$ and $\bar{\psi}$ on M. Since any Hamiltonian vector field is tangent to S by definition, the restriction of the Poisson bracket $\{\varphi, \psi\}$ to S does not depend on the choice of the extensions. This defines an induced bracket operation in $C^\infty(S)$ which is easily seen to make S into a Poisson submanifold of M.

Now we observe that for any point $x \in S$ we have an isomorphism $\check{\pi}_S : T^*S \to TS$. This immediately implies that S is a symplectic manifold. Finally, the statement (3) follows from the definition of symplectic leaf as an equivalence class. □

A natural question to ask about a Poisson manifold is how to describe its symplectic leaves. We are going to state a theorem that provides a useful tool for that.

DEFINITION 1.2.4. Let P_1 and P_2 be Poisson manifolds, S a symplectic manifold. A diagram $P_1 \xleftarrow{f_1} S \xrightarrow{f_2} P_2$ is called *dual pair* if f_1 and f_2 are Poisson maps and the Poisson subalgebras $f_1^*C^\infty(P_1)$ and $f_2^*C^\infty(P_2)$ of $C^\infty(S)$ centralize each other with respect to the Poisson bracket.

A dual pair is called *full* if f_1 and f_2 are submersions.

THEOREM 1.2.5. *Let* $P_1 \xleftarrow{f_1} S \xrightarrow{f_2} P_2$ *be a full dual pair. Then the blow-up* $M_{x_1} = f_2 f_1^{-1}(x_1)$ *is a symplectic leaf in* P_2 *for any* $x_1 \in P_1$.

PROOF. Let $x_2 \in M_{x_1}$ be an arbitrary point, and let $y \in S$ be such that $f_1(y) = x_1$ and $f_2(y) = x_2$. Consider the tangent space $V = T_yS$, which is obviously a symplectic vector space. We denote the symplectic form on V by ω_y. The proof of the theorem consists of several steps.

LEMMA 1.2.6. *The tangent spaces to the fibers* $f_1^{-1}(x_1)$ *and* $f_2^{-1}(x_2)$ *are orthogonal to each other with respect to* ω_y. *Moreover, they are orthogonal complements of each other.*

PROOF. Indeed, let $\varphi_1, \ldots, \varphi_{n_1}$ be local coordinates at x_1, and $\psi_1, \ldots, \psi_{n_2}$ local coordinates at x_2. Then the functions $f_1^*(\varphi_i)$ ($i = 1, \ldots, n_1$) (resp. $f_2^*(\psi_i)$ ($i = 1, \ldots, n_2$)) are constant along the fiber $f_1^{-1}(x_1)$ (resp. $f_2^{-1}(x_2)$). Hence, a vector $v \in T_yS$ is tangent to $f_1^{-1}(x_1)$ (resp. $f_2^{-1}(x_2)$) if and only if $\langle df_1^*(\varphi_i), v \rangle = 0$ ($i = 1, \ldots, n_1$) (resp. $\langle df_2^*(\psi_i), v \rangle = 0$ ($i = 1, \ldots, n_2$)).

Since we have a dual pair, $\xi_{\psi_i}(y)$ ($i = 1, \ldots, n_2$) are tangent to $f_1^{-1}(x_1)$, and $\xi_{\varphi_i}(y)$ ($i = 1, \ldots, n_1$) are tangent to $f_2^{-1}(x_2)$. Moreover, we see that the vector space V_1 generated by the vectors $\xi_{\psi_i}(y)$ is symplectically orthogonal to the vector space V_2 generated by the vectors $\xi_{\varphi_i}(y)$.

On the other hand, let $v \in V$ be a vector tangent to $f_1^{-1}(x_1)$ at y. Since S is symplectic, we may assume that $v = \xi_\varphi(y)$ for some smooth function φ. Thus, we see that $\{\varphi, f_1^*(\varphi_i)\} = 0$ for $i = 1, \ldots, n_1$, hence $\varphi \in f_2^*(C^\infty(P_2))$. Therefore, $\xi_\varphi(y) \in V_1$, and we conclude that V_1 coincides with the tangent space to $f_1^{-1}(x_1)$. Similarly, V_2 coincides with the tangent space to $f_2^{-1}(x_2)$. This also shows that V_1 and V_2 are symplectic orthogonal complements of each other. □

EXERCISE 1.2.7. M_{x_1} is a submanifold in P_2.

EXERCISE 1.2.8. If V is a symplectic vector space, $W \subset V$ an arbitrary vector subspace, and $W^\perp$ the symplectic orthogonal complement to W, then $W/(W \cap W^\perp)$ carries a natural symplectic structure induced by V.

By Lemma 1.2.6, we have that $V_2 = V_1^\perp$. On the other hand, the tangent space $T_{x_2} P_2$ is isomorphic to $V/\operatorname{Ker}(f_2)_* \simeq V/V_2$, and $T_{x_2} M_{x_1}$ is isomorphic to $V_1/(V_1 \cap V_2) \simeq V_1/(V_1 \cap V_1^\perp)$. Then by Exercise 1.2.8, we get a symplectic form η_{x_2} on $T_{x_2} M_{x_1}$. The next exercise completes the proof of the theorem.

EXERCISE 1.2.9. As x_2 varies, the form η_{x_2} defines a symplectic structure on M_{x_1}. It is a Poisson submanifold of P_2.

This completes the proof of Theorem 1.2.5. □

PROPOSITION 1.2.10. *Let A and B be Poisson algebras. The formula*

$$(1.2.1) \qquad \{a \otimes b, c \otimes d\} = \{a, c\} \otimes bd + ac \otimes \{b, d\}$$

makes $A \otimes B$ a Poisson algebra which is called the tensor product *of the Poisson algebras A and B.*

This operation satisfies all the usual properties of tensor product so that the category of Poisson algebras becomes a monoidal category.[1]

PROOF. A straightforward computation checks that all the properties of the Poisson bracket in Definition 1.1.1 are satisfied. □

REMARK 1.2.11. If A and B are Poisson algebras with unity, then the Poisson algebra structure on $A \otimes B$ given by (1.2.1) is the only one which satisfies the following properties:
 (1) the canonical embeddings $A = A \otimes 1 \hookrightarrow A \otimes B$ and $B = 1 \otimes B \hookrightarrow A \otimes B$ are Poisson homomorphisms;
 (2) $\{A \otimes 1, 1 \otimes B\} = 0$.

EXERCISE 1.2.12. Show that for any two Poisson manifolds M and N, there exists a unique Poisson manifold structure on $M \times N$ such that the Poisson algebra structure on $C^\infty(M) \otimes C^\infty(N) \subset C^\infty(M \times N)$ coincides with that of the tensor product of the Poisson algebras.

The Poisson manifold $M \times N$ is called the *Cartesian product* of the Poisson manifolds M and N.

REMARK 1.2.13. Translation of this definition from the language of functions into the language of points is as follows. Given the bivector fields π_M and π_N which correspond to the Poisson manifold structures on M and N respectively, we define

---

[1] The definition of monoidal category can be found in Section 6 of Chapter 2.

the Cartesian product Poisson manifold structure on $M \times N$ by the bivector field

(1.2.2) $$\pi_{M\times N}(m,n) = \pi_M(m) + \pi_N(n),$$

where $m \in M$ and $n \in N$.

EXERCISE 1.2.14. Let M be a Poisson manifold, π the corresponding bivector field, and $\check{\pi} : T^*M \to TM$ the bundle map induced by π.

(1) The space $\Omega^1(M)$ of differential 1-forms on M has a Lie algebra structure, with the Lie bracket given by the formula

(1.2.3) $$\begin{aligned}[\alpha_1, \alpha_2] &= -d\pi(\alpha_1, \alpha_2) + \mathcal{L}_{\check{\pi}\alpha_2}\alpha_1 - \mathcal{L}_{\check{\pi}\alpha_1}\alpha_2 \\ &= d\pi(\alpha_1, \alpha_2) - \check{\pi}\alpha_1 \rfloor d\alpha_2 + \check{\pi}\alpha_2 \rfloor d\alpha_1,\end{aligned}$$

for any $\alpha_1, \alpha_2 \in \Omega^1(M)$, where $\mathcal{L}$ is the symbol of Lie derivative and $\rfloor$ denotes conraction of a differential form with a vector field.

(2) Moreover, $-\check{\pi}$ is a Lie algebra homomorphism of $\Omega^1(M)$ into the Lie algebra $Vect(M)$ of vector fields on M, and the following derivation law holds:

$$[\varphi\alpha_1, \alpha_2] = \varphi[\alpha_1, \alpha_2] - (-\check{\pi}\alpha_2 \cdot \varphi)\alpha_1,$$

for any $\varphi \in C^\infty(M)$, $\alpha_1, \alpha_2 \in \Omega^1(M)$.

2. Lie bialgebras and Manin triples

In the next section we will introduce the notion of Poisson Lie group. Before doing so, we want to discuss its infinitesimal counterpart, the notion of Lie bialgebra, as it can be understood in the purely algebraic context.

Throughout the book we assume that all Lie algebras and Lie groups are finite-dimensional.

2.1. Lie bialgebras.

DEFINITION 2.1.1. A *Lie bialgebra* is a pair $(\mathfrak{g}, \varphi)$ where $\mathfrak{g}$ is a Lie algebra over **k**, and $\varphi : \mathfrak{g} \to \mathfrak{g} \wedge \mathfrak{g}$ (called *cobracket*) satisfies the following conditions:

(1) the dual map $\varphi^* : \mathfrak{g}^* \wedge \mathfrak{g}^* \to \mathfrak{g}^*$ makes $\mathfrak{g}^*$ a Lie algebra;

(2) the cobracket $\varphi : \mathfrak{g} \to \mathfrak{g} \wedge \mathfrak{g}$ is a 1-cocycle on $\mathfrak{g}$, the $\mathfrak{g}$-module structure on $\mathfrak{g} \wedge \mathfrak{g}$ given by the adjoint action. In other words, we have that for any $a, b \in \mathfrak{g}$,

(2.1.1) $$\varphi([a,b]) = a.\varphi(b) - b.\varphi(a),$$

where

$$a.(b \otimes c) = [a \otimes 1 + 1 \otimes a, b \otimes c] = [a,b] \otimes c + b \otimes [a,c].$$

EXERCISE 2.1.2. Check that one can give an equivalent definition of Lie bialgebra by considering a pair $(\mathfrak{g}, \mathfrak{g}^*)$ of Lie algebras which are dual to each other as vector spaces and their Lie algebra structures are compatible in the following way:

(2.1.2) $$\sum_k c^k_{rs} f^{ij}_k = \sum_p \left(c^i_{pr} f^{jp}_s - c^j_{pr} f^{ip}_s - c^i_{ps} f^{jp}_r + c^j_{ps} f^{ip}_r \right)$$

(for any $i, j, r, s = 1, \ldots, \dim \mathfrak{g}$), where $c^\gamma_{\alpha\beta}$ and $f^{\alpha\beta}_\gamma$ are the structure constants of $\mathfrak{g}$ and $\mathfrak{g}^*$, respectively: given some basis $\{I_\alpha\}$ of $\mathfrak{g}$ and the dual basis $\{I^\alpha\}$ of $\mathfrak{g}^*$, one has

$$[I_\alpha, I_\beta] = \sum_\gamma c^\gamma_{\alpha\beta} I_\gamma, \quad [I^\alpha, I^\beta] = \sum_\gamma f^{\alpha\beta}_\gamma I^\gamma.$$

REMARK 2.1.3. The Lie bialgebras form a category, with the morphisms being Lie algebra homomorphisms which commute with the cobrackets.

PROPOSITION 2.1.4. *Given a Lie bialgebra* $\mathfrak{g}$, *the vector space* $\mathfrak{g}^*$ *carries a canonical Lie bialgebra structure, the Lie bracket being the map dual to the cobracket in* $\mathfrak{g}$ *and the cobracket being the map dual to the bracket in* $\mathfrak{g}$.

PROOF. By definition, $\mathfrak{g}^*$ is a Lie algebra. Denote by $\langle\,,\,\rangle : \mathfrak{g} \otimes \mathfrak{g}^* \to \mathbf{k}$ the natural pairing between the dual spaces $\mathfrak{g}$ and $\mathfrak{g}^*$. Let $\varphi : \mathfrak{g} \to \mathfrak{g} \wedge \mathfrak{g}$ be the cobracket in $\mathfrak{g}$, and $\varphi_* : \mathfrak{g}^* \to \mathfrak{g}^* \wedge \mathfrak{g}^*$ the linear map dual to the Lie bracket in $\mathfrak{g}$. Then for any $a, b \in \mathfrak{g}$ and $\alpha, \beta \in \mathfrak{g}^*$, we have

$$\begin{aligned}
\langle a \otimes b, \varphi_*([\alpha, \beta])\rangle &= \langle [a,b], [\alpha, \beta]\rangle = \langle \varphi([a,b]), \alpha \otimes \beta\rangle \\
&= \langle [a \otimes 1 + 1 \otimes a, \varphi(b)] - [b \otimes 1 + 1 \otimes b, \varphi(a)], \alpha \otimes \beta\rangle \\
&= \langle a \otimes \varphi(b), \varphi_*(\alpha) \otimes \beta\rangle + \langle P^{12}(a \otimes \varphi(b)), \alpha \otimes \varphi_*(\beta)\rangle \\
&\quad - \langle b \otimes \varphi(a), \varphi(\alpha) \otimes \beta\rangle - \langle P^{23}(\varphi(a) \otimes b), \alpha \otimes \varphi_*(\beta)\rangle \\
&= \langle a \otimes b, [\alpha \otimes 1 + 1 \otimes \alpha, \varphi_*(\beta)] - [\beta \otimes 1 + 1 \otimes \beta, \varphi_*(\alpha)]\rangle,
\end{aligned}$$

where $P^{12} : x \otimes y \otimes z \mapsto y \otimes x \otimes z$ and $P^{23} : x \otimes y \otimes z \mapsto x \otimes z \otimes y$ are the permutations in $\mathfrak{g}^{\otimes 3}$. This proves the proposition. □

DEFINITION 2.1.5. The Lie bialgebra $\mathfrak{g}^*$ is called the dual Lie bialgebra of $\mathfrak{g}$.

EXAMPLE 2.1.6. It is easy to see that any Lie algebra $\mathfrak{g}$ equipped with the trivial (i.e., zero) cobracket satisfies all the axioms of Lie bialgebra.

EXAMPLE 2.1.7. Let $\mathfrak{g}$ be a Lie algebra. Consider the dual space $\mathfrak{g}^*$ as an abelian Lie algebra. Then the map $\varphi : \mathfrak{g}^* \to \mathfrak{g}^* \wedge \mathfrak{g}^*$ dual to the Lie bracket on $\mathfrak{g}$, defines a Lie bialgebra structure on $\mathfrak{g}^*$.

It is easy to see that this is the dual Lie bialgebra to the one described in Example 2.1.6.

EXAMPLE 2.1.8. Consider the Lie algebra $\mathfrak{g} = \mathfrak{sl}(2, \mathbb{C})$ of traceless complex 2×2 matrices. Recall that it is generated by

$$X^+ = \begin{pmatrix} 0 & 1 \\ 0 & 0 \end{pmatrix}, \quad X^- = \begin{pmatrix} 0 & 0 \\ 1 & 0 \end{pmatrix}, \quad H = \begin{pmatrix} 1 & 0 \\ 0 & -1 \end{pmatrix}$$

and the relations

$$[H, X^\pm] = \pm 2 X^\pm, \quad [X^+, X^-] = H.$$

The following cobracket makes $\mathfrak{sl}(2, \mathbb{C})$ a Lie bialgebra:

$$\varphi\left(X^\pm\right) = X^\pm \wedge H, \quad \varphi(H) = 0.$$

EXAMPLE 2.1.9. Let $\mathfrak{g}$ be a complex simple Lie algebra with a fixed nondegenerate invariant symmetric bilinear form. Suppose that we have chosen a Cartan subalgebra $\mathfrak{h}$ of $\mathfrak{g}$, and $n = \dim \mathfrak{h}$ is the rank of $\mathfrak{g}$. Suppose that we have chosen also a set of simple roots $\alpha_1, \ldots, \alpha_n \in \mathfrak{h}^*$. This gives a decomposition

(2.1.3) $$\mathfrak{g} = \mathfrak{n}_+ \oplus \mathfrak{h} \oplus \mathfrak{n}_-$$

where $\mathfrak{n}_+$ (resp. $\mathfrak{n}_-$) is the nilpotent Lie subalgebra of $\mathfrak{g}$ spanned by the positive (resp. negative) root subspaces.

Let $X_i^\pm, H_i$ ($i = 1, \ldots, n$) be the Chevalley generators corresponding to the simple roots α_i, and $A = (a_{ij})_{i,j=1}^n$ the Cartan matrix whose entries are $a_{ij} = $

$\frac{2(\alpha_i,\alpha_j)}{(\alpha_i,\alpha_i)}$, where (,) is the symmetric bilinear form on $\mathfrak{h}^*$ induced by the bilinear form on $\mathfrak{g}$.

Recall that $\mathfrak{g}$ is generated by $X_i^\pm, H_i$ and the relations

(2.1.4) $\qquad\qquad [H_i, H_j] = 0, \qquad [H_i, X_j^\pm] = \pm a_{ij} X_j^\pm,$

(2.1.5) $\qquad\qquad [X_i^+, X_j^-] = \delta_{ij} H_i, \qquad \mathrm{ad}^{1-a_{ij}}\left(X_i^\pm\right) X_j^\pm = 0,$

where $\mathrm{ad}(a)b = [a,b]$ is the adjoint action of $\mathfrak{g}$ on itself.

We leave it as an exercise for the reader to check that the following cobracket φ defines a Lie bialgebra structure on $\mathfrak{g}$:

(2.1.6) $\qquad\qquad \varphi(H_i) = 0, \quad \varphi\left(X_i^\pm\right) = d_i X_i^\pm \wedge H_i,$

where d_i ($i = 1, \ldots, n$) are positive rational numbers that satisfy $d_i a_{ij} = d_j a_{ji}$.

DEFINITION 2.1.10. The Lie bialgebra structure on $\mathfrak{g}$ described in Example 2.1.9 is called the *standard* Lie bialgebra structure on $\mathfrak{g}$.

In Section 4 we will observe that there are many other Lie bialgebra structures on a complex simple Lie algebra $\mathfrak{g}$. They are partially classified by Belavin–Drinfeld Theorem 4.2.2.

EXERCISE 2.1.11. Show that all standard Lie bialgebra structures on a complex simple Lie algebra $\mathfrak{g}$ are equivalent up to conjugation and hence are independent of the choice of the decomposition (2.1.3).

2.2. Co-Poisson Hopf algebras. Suppose that $\mathfrak{g}$ is a Lie algebra over a field $\mathbf{k}$. Then the universal enveloping algebra $U\mathfrak{g}$ carries a canonical Hopf algebra structure. It turns out that if $\mathfrak{g}$ is equipped with a Lie bialgebra structure, $U\mathfrak{g}$ becomes a co-Poisson Hopf algebra.

First let us recall the definiton of Hopf algebra. Note that we give a rather special definition, as sometimes people do not require that there be a unity, an antipode, and a counit, or they do not require that the antipode be invertible.

DEFINITION 2.2.1. A *Hopf algebra* over a field $\mathbf{k}$ is a quadruple $(A, \Delta, \varepsilon, S)$ where A is an associative algebra with unity, $\Delta : A \to A \otimes A$ is an algebra homomorphism called *comultiplication*, $\varepsilon : A \to \mathbf{k}$ an algebra homomorphism called *counit*, and $S : A \to A$ an algebra antiautomorphism called *antipode* such that the following diagrams commute:

(2.2.1)
$$\begin{array}{ccc} & A \otimes A & \\ {}^{\Delta}\nearrow & & \searrow{}^{\Delta \otimes \mathrm{id}} \\ A & & A \otimes A \otimes A \\ {}_{\Delta}\searrow & & \nearrow{}_{\mathrm{id} \otimes \Delta} \\ & A \otimes A & \end{array} \qquad \begin{array}{ccc} A & \xrightarrow{\Delta} & A \otimes A \\ \Delta \downarrow & \searrow \mathrm{id} & \downarrow \varepsilon \otimes \mathrm{id} \\ A \otimes A & \xrightarrow{\mathrm{id} \otimes \varepsilon} & A \end{array}$$

(coassociativity) $\qquad\qquad\qquad$ (counit condition)

(2.2.2)
$$\begin{array}{ccccc} A \otimes A & \xleftarrow{\Delta} & A & \xrightarrow{\Delta} & A \otimes A \\ S \otimes \mathrm{id} \downarrow & & \varepsilon \cdot \mathrm{id} \downarrow & & \downarrow \mathrm{id} \otimes S \\ A \otimes A & \xrightarrow{m} & A & \xleftarrow{m} & A \otimes A \end{array}$$

(antipode condition)

where

(2.2.3) $\qquad\qquad \Delta(S(a)) = (S \otimes S)\Delta'(a)$

for any $a \in A$, where $m : A \otimes A \to A$ is the multiplication map,

$$\Delta' = \sigma\Delta \quad (\text{where } \sigma : a \otimes b \mapsto b \otimes a),$$

for any $a, b \in A$ (Δ' is called *opposite comultiplication*).

REMARK 2.2.2. If we forget about the algebra structure on a Hopf algebra A, we get what is called *coalgebra*. The condition (2.2.3) means that the antipode S is a coalgebra antihomomorphism.

EXAMPLE 2.2.3. There is a canonical Hopf algebra structure on the universal enveloping algebra $U\mathfrak{g}$ of a Lie algebra $\mathfrak{g}$. The comultiplication Δ, the antipode S and the counit ε are defined on the generators $a \in \mathfrak{g} \subset U\mathfrak{g}$ as follows:

(2.2.4) $$\Delta(a) = a \otimes 1 + 1 \otimes a,$$
(2.2.5) $$S(a) = -a, \quad \varepsilon(a) = 0.$$

Now we recall the notion of the dual Hopf algebra. Although in this book we deal with infinite-dimensional Hopf algebras only (as in Example 2.2.3, for instance), let us discuss the finite-dimensional case first.

PROPOSITION 2.2.4. *Suppose that A is a finite-dimensional Hopf algebra over a field $\mathbf{k}$ of characteristic zero. There is a canonical Hopf algebra structure on the dual space A^* given by*

(2.2.6) $$\langle a, f_1 f_2 \rangle = \langle \Delta(a), f_1 \otimes f_2 \rangle,$$

(2.2.7) $$\langle a \otimes b, \Delta(f) \rangle = \langle ab, f \rangle,$$

(2.2.8) $$\langle a, S(f) \rangle = \langle S(a), f \rangle, \quad \varepsilon(f) = \langle f, 1 \rangle,$$

where $a, b \in A$, $f, f_1, f_2 \in A^*$ and $\langle \, , \, \rangle : A \otimes A^* \to \mathbf{k}$ is the natural pairing.

PROOF. It is a straightforward computation to check that all the axioms of Hopf algebra are satisfied. $\square$

REMARK 2.2.5. When A is infinite-dimensional, A^* is still an algebra, but formulas (2.2.7), (2.2.8) do not define a coalgebra structure in general, since Δ acts from A to $(A \otimes A)^*$ and not necessarily to $A^* \otimes A^* \subset (A \otimes A)^*$.

However, one can choose a subalgebra $A^\bullet \subset A^*$ which is a Hopf algebra with respect to the structure given by (2.2.6)–(2.2.8).

PROPOSITION 2.2.6. *Suppose that A is a Hopf algebra with unity over a field $\mathbf{k}$, and A^* the dual space. Let $A^\bullet$ be a subspace of A^* such that for any $f \in A^\bullet$ there exists a two-sided ideal $J \subset A$ of finite codimension annihilated by f.*

Then $A^\bullet$ is a Hopf algebra, with the structure given by (2.2.6)–(2.2.8). It is called dual Hopf algebra to A, *or the* restricted dual to A.

PROOF. We prove first that $A^\bullet$ is an algebra. Let $\langle \, , \, \rangle : A \otimes A^* \to \mathbf{k}$ be the natural pairing. There is a natural A-bimodule structure on A^* given by the actions $a : f \mapsto a.f$ and $a : f \mapsto f.a$ given by

$$\langle b, a.f \rangle = \langle ba, f \rangle,$$
$$\langle b, f.a \rangle = \langle ab, f \rangle,$$

for any $a, b \in A$ and $f \in A^*$.

For any $f \in A^\bullet$, consider the linear subspace $A \cdot f \cdot A \subset A^*$. Let J_f be the maximal two-sided ideal in A annihilated by $A \cdot f \cdot A$. Obviously, it is annihilated by f and, thus, has a finite codimension in A. It is easy to see that $(A \cdot f \cdot A)(A \cdot g \cdot A) = A \cdot fg \cdot A$ for any $f, g \in A^\bullet$. This implies that $\Delta(J_{fg}) \subset J_f \otimes A + A \otimes J_g$ has a finite codimension in $A \otimes A$. On the other hand, if $\Delta(a) \in J_f \otimes A + A \otimes J_g$, then $\langle a, fg \rangle = \langle \Delta(a), f \otimes g \rangle = 0$, and $a \in J_{fg}$. Therefore, J_{fg} also has a finite codimentsion in A, which proves that $fg \in A^\bullet$.

To prove that the linear map Δ in A^* dual to the multiplication map in A defines a comultiplication in $A^\bullet$, note that if J is a two-sided ideal in A of finite codimension which is annihilated by $f \in A^\bullet$, then $J \otimes A + A \otimes J$ is a two-sided ideal of finite codimension annihilated by $\Delta(f)$. We leave it to the reader to prove that the counit and the antipode are well defined in $A^\bullet$. □

DEFINITION 2.2.7. Let A be a Hopf algebra. Then the Hopf algebra $A^\bullet$ defined in Proposition 2.2.6 is called the *dual Hopf algebra*.

EXAMPLE 2.2.8. Let G be an affine connected algebraic group over a field $\mathbf{k}$, and $\mathfrak{g}$ the Lie algebra of G. Then the algebra $\mathbf{k}[G]$ of regular functions on G is the Hopf algebra dual to the universal enveloping algebra $U\mathfrak{g}$ with respect to the pairing $\langle \, , \, \rangle : U\mathfrak{g} \otimes \mathbf{k}[G] \to \mathbf{k}$ given by

$$\langle \xi, f \rangle = \mathcal{R}(\xi)f|_{g=e} = \mathcal{L}(\xi)f|_{g=e}$$

where $\mathcal{R}$ (resp. $\mathcal{L}$) is the canonical representation of $U\mathfrak{g}$ as the algebra of right (resp. left) invariant differential operators in $\mathbf{k}[G]$, and e is the unit element of G.

DEFINITION 2.2.9. A *co-Poisson Hopf algebra* is a pair (U, δ) where U is a Hopf algebra and the linear map $\delta : U \to U \otimes U$ (called *cobracket*) is such that

(2.2.9) $$\delta(ab) = \Delta(a)\delta(b) + \delta(a)\Delta(b)$$

and the dual map $\delta^* : U^* \otimes U^* \to U^*$ makes U^* a Poisson algebra with the Poisson bracket δ^*.

PROPOSITION 2.2.10. *Let $(\mathfrak{g}, \varphi)$ be a Lie bialgebra and $U = U\mathfrak{g}$ the universal enveloping algebra. There exists a unique cobracket $\hat{\varphi} : U\mathfrak{g} \to U\mathfrak{g} \otimes U\mathfrak{g}$ such that $\delta|_{\mathfrak{g}} = \varphi$.*

In particular, $U\mathfrak{g}$ has a canonical co-Poisson Hopf algebra structure.

PROOF. Define δ by (2.2.9) and $\delta|_{\mathfrak{g}} = \varphi$. The correctness of the definition follows immediately from

$$\delta([a, b]) = [\Delta(a), \delta(b)] + [\delta(a), \Delta(b)],$$

for any $a, b \in \mathfrak{g}$, which is equivalent to the property that $\delta|_{\mathfrak{g}} = \varphi$ is a 1-cocycle on $\mathfrak{g}$. □

2.3. Manin triples. It is often convenient to think of Lie bialgebras in terms of Manin triples.

DEFINITION 2.3.1. Let $\mathfrak{g}$ be a Lie algebra equipped with a nondegenerate invariant symmetric bilinear form $\langle , \rangle$, and $\mathfrak{g}_+$ and $\mathfrak{g}_-$ Lie subalgebras of $\mathfrak{g}$ such that

$$\mathfrak{g} = \mathfrak{g}_+ \oplus \mathfrak{g}_-$$

as a vector space. The triple $(\mathfrak{g}, \mathfrak{g}_+, \mathfrak{g}_-)$ is called a *Manin triple* if both $\mathfrak{g}_+$ and $\mathfrak{g}_-$ are maximal isotropic subspaces of $\mathfrak{g}$ with respect to the form $\langle , \rangle$.

Suppose that $(\mathfrak{g}, \mathfrak{g}_+, \mathfrak{g}_-)$ is a Manin triple. Since both $\mathfrak{g}_+$ and $\mathfrak{g}_-$ are maximal isotropic subspaces of $\mathfrak{g}$, we can identify $\mathfrak{g}_\mp$ with $\mathfrak{g}_\pm^*$ as a vector space.

THEOREM 2.3.2. (1) *Suppose that $(\mathfrak{g}, \mathfrak{g}_+, \mathfrak{g}_-)$ is a Manin triple. Then $(\mathfrak{g}_\pm, \varphi)$ is a Lie bialgebra where the cobracket $\varphi : \mathfrak{g}_\pm \to \mathfrak{g}_\pm \wedge \mathfrak{g}_\pm$ is the dual map to the Lie bracket in $\mathfrak{g}_\mp$.*

(2) *Suppose that $(\mathfrak{g}, \varphi)$ is a Lie bialgebra. Consider the vector space $\mathfrak{g}^*$ equipped with the dual Lie bialgebra structure (cf. Definition 2.1.5). Then*
$$(\mathfrak{g} \oplus \mathfrak{g}^*, \mathfrak{g}, \mathfrak{g}^*)$$
is a Manin triple, with the Lie algebra structure on $\mathfrak{g} \oplus \mathfrak{g}^*$ given by
(2.3.1)
$$[a + \alpha, b + \beta] = [a, b] + [\alpha, \beta] + \mathrm{ad}^*_\alpha b - \mathrm{ad}^*_\beta a - \mathrm{ad}^*_b \alpha + \mathrm{ad}^*_a \beta,$$
for any $a, b \in \mathfrak{g}$, $\alpha, \beta \in \mathfrak{g}^*$, and the bilinear form on $\mathfrak{g} \oplus \mathfrak{g}^*$ given by
$$\langle (a, \alpha), (b, \beta) \rangle = \beta(a) + \alpha(b).$$

PROOF. Given a Manin triple $(\mathfrak{g}, \mathfrak{g}_+, \mathfrak{g}_-)$, define a cobracket $\varphi : \mathfrak{g}_\pm \to \mathfrak{g}_\pm \wedge \mathfrak{g}_\pm$ as the linear map dual to the Lie bracket in $\mathfrak{g}_\mp$. Thus, the property (1) of Definition 2.1.1 is fulfilled automatically.

It follows from the invariance of the bilinear form $\langle \, , \, \rangle$ that
$$\begin{aligned} [\alpha, b] &= \mathrm{ad}^*_\alpha b - \mathrm{ad}^*_b \alpha, \\ [a, \beta] &= \mathrm{ad}^*_a \beta - \mathrm{ad}^*_\beta a, \end{aligned}$$
for any $a, b \in \mathfrak{g}_\pm$ and $\alpha, \beta \in \mathfrak{g}_\mp$. Therefore,
$$\begin{aligned} (\varphi([a, b]), \alpha \otimes \beta) &= ([a, b], [\alpha, \beta]) = -(a, [b, [\alpha, \beta]]) = (a, [[\alpha, b], \beta] + [\alpha, [\beta, b]]) \\ &= (a, [\mathrm{ad}^*_\alpha b - \mathrm{ad}^*_b \alpha, \beta] + [\alpha, \mathrm{ad}^*_\beta b - \mathrm{ad}^*_b \beta]) \\ &= ([a, \mathrm{ad}^*_\beta b, \alpha], \alpha) + ([\mathrm{ad}^*_\alpha b, a], \beta) \\ &\quad - (\varphi(a), \alpha \otimes \mathrm{ad}^*_b \beta) - (\varphi(a), \mathrm{ad}^*_b \alpha \otimes \beta) \\ &= ([a \otimes 1 + 1 \otimes a, \varphi(b)] - [b \otimes 1 + 1 \otimes b, \varphi(a)], \alpha \otimes \beta), \end{aligned}$$
which proves the cocyclicity of φ. Thus, the first statement of the theorem is proved.

We leave it as an exercise to the reader to check that the bracket (2.3.1) satisfies the Jacobi identity. Therefore, $(\mathfrak{g} \oplus \mathfrak{g}^*, \mathfrak{g}, \mathfrak{g}^*)$ is a Manin triple. □

PROPOSITION 2.3.3. *Let $\mathfrak{g}$ be a Lie bialgebra, and $\mathcal{D}(\mathfrak{g}) = \mathfrak{g} \oplus \mathfrak{g}^*$ the Lie algebra described in Theorem 2.3.2. Then there exists a canonical Lie bialgebra structure on $\mathcal{D}(\mathfrak{g})$ given by the cobracket*
$$\varphi_{\mathcal{D}(\mathfrak{g})}(a + \alpha) = \varphi_\mathfrak{g}(a) + \varphi_{\mathfrak{g}^*}(\alpha),$$
for any $a \in \mathfrak{g}$, $\alpha \in \mathfrak{g}^*$.

PROOF. Follows immediately from (2.3.1). □

DEFINITION 2.3.4. The Lie bialgebra $\mathcal{D}(\mathfrak{g})$ is called *the double Lie bialgebra* of $\mathfrak{g}$ (sometimes it is called *the classical double* of $\mathfrak{g}$).

REMARK 2.3.5. The Manin triple which corresponds to the double Lie bialgebra is
$$(\mathcal{D}(\mathfrak{g}) \oplus \mathcal{D}(\mathfrak{g}), \mathcal{D}(\mathfrak{g}), \mathfrak{g} \oplus \mathfrak{g}^*)$$

where $\mathcal{D}(\mathfrak{g})$ is embedded into $\mathcal{D}(\mathfrak{g}) \oplus \mathcal{D}(\mathfrak{g})$ as the diagonal, and $\mathfrak{g} \oplus \mathfrak{g}^*$ (the Lie bialgebra dual to $\mathcal{D}(\mathfrak{g})$) is the direct sum of the Lie algebras $\mathfrak{g} = (\mathfrak{g}, 0) \subset \mathcal{D}(\mathfrak{g}) \oplus \mathcal{D}(\mathfrak{g})$ and $\mathfrak{g}^* = (0, \mathfrak{g}^*) \subset \mathcal{D}(\mathfrak{g}) \oplus \mathcal{D}(\mathfrak{g})$. The bilinear form is given by

$$\langle (a,b), (c,d) \rangle = \langle a, c \rangle - \langle b, d \rangle \tag{2.3.2}$$

for any $a, b, c, d \in \mathcal{D}(\mathfrak{g})$. Here $\langle \, , \, \rangle$ on the right-hand side of (2.3.2) denotes the bilinear form on $\mathcal{D}(\mathfrak{g})$.

The double Lie bialgebras have an important property. Namely, the cobracket on such a Lie bialgebra is described by a very simple formula, as shown in the following proposition.

PROPOSITION 2.3.6. *Let $\mathfrak{g}$ be a Lie bialgebra, and $\mathcal{D}(\mathfrak{g})$ the double Lie bialgebra. Suppose that $\{I_\alpha\}$ is a basis in $\mathfrak{g}$ and $\{I^\alpha\}$ the dual basis of $\mathfrak{g}^*$. Let us identify I_α with $(I_\alpha, 0) \in \mathcal{D}(\mathfrak{g})$ and I^α with $(0, I^\alpha) \in \mathcal{D}(\mathfrak{g})$. Then the cobracket in $\mathcal{D}(\mathfrak{g})$ is given by*

$$\varphi(a) = [a \otimes 1 + 1 \otimes a, r] \tag{2.3.3}$$

for any $a \in \mathcal{D}(\mathfrak{g})$, where

$$r = \sum_\alpha I_\alpha \otimes I^\alpha \in \mathcal{D}(\mathfrak{g}) \otimes \mathcal{D}(\mathfrak{g}).$$

PROOF. The property (2.1.1) is equivalent to the property of the adjoint action of $\mathfrak{g}$ in $\mathfrak{g} \otimes \mathfrak{g}$ to be a Lie algebra representation, since $\varphi(a) = a \cdot r$. The fact that the dual map φ^* satisfies the Jacobi identity can be checked by a straightforward computation. We will see later that this is just a special case of Proposition 4.1.1, and the Jacobi identity for φ^* is equivalent to (4.1.2). □

The following list of exercises provides some important examples of Lie bialgebras and corresponding Manin triples.

EXERCISE 2.3.7. Let $\mathfrak{g}$ be a complex simple Lie algebra equipped with the standard Lie bialgebra structure (cf. Definition 2.1.10) with a fixed invariant bilinear form $\langle \, , \, \rangle$. Show that the corresponding Manin triple is $(\mathfrak{g} \oplus \mathfrak{g}, \mathfrak{g}, \mathfrak{s})$, where $\mathfrak{g}$ is embedded diagonally into $\mathfrak{g} \oplus \mathfrak{g}$ and $\mathfrak{s} = \{(x, y) \in \mathfrak{b}_+ \oplus \mathfrak{b}_- \mid x_\mathfrak{h} + y_\mathfrak{h} = 0\}$. Here $a_\mathfrak{h}$ denotes the "Cartan part" of $a \in \mathfrak{b}_\pm$. In other words, $a = a_\pm + a_\mathfrak{h}$, where $a_\pm \in \mathfrak{n}_\pm$ and $a_\mathfrak{h} \in \mathfrak{h}$. The invariant bilinear form on $\mathfrak{g} \oplus \mathfrak{g}$ is given by

$$\langle (a,b), (c,d) \rangle = \langle a, c \rangle - \langle b, d \rangle.$$

Show that the dual Lie bialgebra has the same vector space decomposition $\mathfrak{g}^* \simeq \mathfrak{n}_+ \oplus \mathfrak{h} \oplus \mathfrak{n}_-$ as $\mathfrak{g}$ does, with the only difference that the Lie subalgebras $\mathfrak{n}_+$ and $\mathfrak{n}_-$ commute. The double Lie bialgebra is isomorphic to $\mathfrak{g} \oplus \mathfrak{g}$ as a Lie algebra.

EXERCISE 2.3.8. Let $\mathfrak{g}$ be as in the previous example. Consider the Borel subalgebras $\mathfrak{b}_\pm = \mathfrak{n}_\pm \oplus \mathfrak{h} \subset \mathfrak{g}$. Show that they are in fact Lie bialgebras with respect to the restriction of the cobracket from $\mathfrak{g}$ to $\mathfrak{b}_\pm$.

Show that the corresponding Manin triple is

$$(\mathfrak{g} \oplus \mathfrak{h}, \mathfrak{b}_+, \mathfrak{b}_-)$$

where the Borel subalgebras are embedded into $\mathfrak{g} \oplus \mathfrak{h}$ so that $a \mapsto (a_\pm, \pm a_\mathfrak{h})$, the bilinear form given by

$$\langle (a,b), (c,d) \rangle = \langle a, c \rangle - \langle b, d \rangle$$

with $a, c \in \mathfrak{g}$ and $b, d \in \mathfrak{h}$.

In particular, we see that the Borel subalgebras $\mathfrak{b}_+$ and $\mathfrak{b}_-$ are dual to each other as Lie bialgebras.

EXERCISE 2.3.9. Let $\mathfrak{g}$ be a comlex simple Lie algebra, $\mathfrak{g}_0$ a real form of $\mathfrak{g}$. (Recall that a *real form* of $\mathfrak{g}$ is a real Lie subalgebra $\mathfrak{g}_0 \subset \mathfrak{g}$ such that $\mathfrak{g}$ is isomorphic to the complexification of $\mathfrak{g}_0$.)

Suppose that we have fixed a decomposition (2.1.3) of $\mathfrak{g}$. Show that

$$(\mathfrak{g}_\mathbb{R}, \mathfrak{g}_0, \mathfrak{n}_+ \oplus i\mathfrak{t})$$

is a Manin triple where $\mathfrak{t} = \mathfrak{h} \cap \mathfrak{g}_0$ and $\mathfrak{g}_\mathbb{R}$ is $\mathfrak{g}$ regarded as a real Lie algebra, with the bilinear form being the imaginary part of the Killing form on $\mathfrak{g}$.

The corresponding Lie bialgebra structure on $\mathfrak{g}_0$ is called *standard*.

REMARK 2.3.10. (1) One can show that there is only one standard Lie bialgebra structure on a compact real form $\mathfrak{g}_0$ of $\mathfrak{g}$ up to conjugation, while a standard Lie bialgebra structure on a noncompact real form may depend on the decomposition (2.1.3) in such a way that different decompositions yield nonequivalent Lie bialgebra structures on $\mathfrak{g}_0$.

(2) Note that for different real forms of $\mathfrak{g}$, the dual Lie bialgebras are isomorphic as Lie algebras (but not as Lie bialgebras).

EXERCISE 2.3.11. Consider the real form $\mathfrak{su}(p,q)$ of $\mathfrak{sl}(n,\mathbb{C})$, where $n = p+q$. Equip $\mathfrak{su}(m,n)$ with a standard Lie bialgebra structure. Then the dual Lie bialgebra is isomorphic to the Lie algebra of traceless upper-triangular complex $n \times n$ matrices with positive diagonal entries.

3. Poisson groups

In this section we introduce the basic notion of Poisson Lie group—the global counterpart of that of Lie bialgebra. Following the same general line, we give a definition first in the language of functions to reinterpret it afterwards in the language of points. We start with the case of Poisson algebraic groups. We will consider only affine algebraic groups, omitting the word "affine."

3.1. Poisson affine algebraic groups and Poisson Hopf algebras.

PROPOSITION 3.1.1. *Let G be an affine algebraic group over a field $\mathbf{k}$. There is a canonical Hopf algebra structure on the algebra $\mathbf{k}[G]$ of regular functions on G given by*

(3.1.1) $$(\Delta f)(g_1, g_2) = f(g_1 g_2),$$
(3.1.2) $$(Sf)(g) = f(g^{-1}), \quad \varepsilon(f) = f(e),$$

for any $f \in \mathbf{k}[G]$ and $g, g_1, g_2 \in G$, $e \in G$ being the unit element.

PROOF. There is a contravariant functor from the category of algebraic groups to the category of $\mathbf{k}$-algebras which assigns to a group the algebra of regular functions on it. The Hopf algebra axioms are just dual to the axioms defining the group structure.

In particular, the coassociativity condition (2.2.1) can be obtained by reversing the arrows in the dual diagram expressing the associativity property of the multiplication $m : G \times G \to G$:

$$\begin{array}{ccc}
 & G \times G & \\
{\scriptstyle m}\swarrow & & \nwarrow{\scriptstyle m\times \mathrm{id}} \\
G & & G \times G \times G \\
{\scriptstyle m}\nwarrow & & \swarrow{\scriptstyle \mathrm{id}\times m} \\
 & G \times G &
\end{array}$$

Similarly, the counit and antipode conditions are direct consequences of the existence of the unit element in the group and the existence of the inverse element. $\square$

REMARK 3.1.2. The Hopf algebra structure on $\mathbf{k}[G]$ described in Proposition 3.1.1 coincides with the one which is dual to the Hopf algebra structure on the universal enveloping algebra $U\mathfrak{g}$ of the Lie algebra $\mathfrak{g}$ of G (cf. Example 2.2.8).

DEFINITION 3.1.3. Suppose that a Hopf algebra A is equipped with a Poisson algebra structure. We say that A is a *Poisson Hopf algebra* if both structures are compatible in the sense that the comultiplication $\Delta : A \to A \otimes A$ is a Poisson algebra homomorphism.

REMARK 3.1.4. One can show that the counit $\varepsilon : A \to \mathbf{k}$ is a Poisson algebra homomorphism, and the antipode $S : A \to A$ is a Poisson algebra antiautomorphism.

PROPOSITION 3.1.5. (1) *Let U be a co-Poisson Hopf algebra. Then the dual Hopf algebra $U^\bullet$ is a Poisson Hopf algebra.*
(2) *Let A be a Poisson Hopf algebra. Then the dual Hopf algebra $A^\bullet$ is a co-Poisson Hopf algebra.*

PROOF. Given a co-Poisson Hopf algebra U with a cobracket δ, define a bracket $\{\,,\,\}$ in $U^\bullet$ as the linear map dual to δ. It makes $U^\bullet$ a Poisson algebra, by Definition 2.2.9. On the other hand, (2.2.9) implies that the comultiplication $\Delta : U^\bullet \to U^\bullet \otimes U^\bullet$ is a Poisson algebra homomorphism. The second statement can be proved similarly. $\square$

We know from Proposition 3.1.1 that the algebra $\mathbf{k}[G]$ of regular functions on an algebraic group G is a Hopf algebra. Recall that it is also a Poisson algebra if G is equipped with a structure of algebraic Poisson variety.

DEFINITION 3.1.6. (1) Suppose G is an algebraic group over a field $\mathbf{k}$ equipped with a Poisson manifold structure. We say that G is a *Poisson algebraic group* if the algebra $\mathbf{k}[G]$ of regular functions on G is a Poisson Hopf algebra.
(2) An equivalent definition in the language of points says that G is a *Poisson algebraic group* if the multiplication $m : G \times G \to G$ is a Poisson map.

REMARK 3.1.7. One can show that this implies also that the taking-the-inverse map $s : G \to G$ is an anti-Poisson map, i.e. the dual map $s^* : \mathbf{k}[G] \to \mathbf{k}[G]$ is an antihomomorphism of Poisson algebras.

3.2. Poisson Lie groups. The definition of Poisson algebraic group formulated in the language of points can be easily carried over to the case of Poisson Lie groups. The principal difficulty is that the comultiplication maps $C^\infty(G)$ into $C^\infty(G \times G)$ but not necessarily into $C^\infty(G) \otimes C^\infty(G) \not\simeq C^\infty(G \times G)$.

DEFINITION 3.2.1. Let G be a Lie group and, at the same time, a Poisson manifold. We say that G is a *Poisson Lie group* if the multiplication $m : G \times G \to G$ is a Poisson map.

REMARK 3.2.2. The Poisson Lie groups form a category, with the morphisms being homomorphisms which are at the same time Poisson maps.

PROPOSITION 3.2.3. *Let G be both a Lie group and a Poisson manifold, and let π be the bivector field corresponding to the Poisson manifold structure on G. Then G is a Poisson Lie group if and only if π satisfies the following condition:*

$$(3.2.1) \qquad \pi(g_1 g_2) = (l_{g_1})_* \pi(g_2) + (r_{g_2})_* \pi(g_1),$$

for any $g_1, g_2 \in G$, where l_{g_i} (resp. r_{g_i}) is the map $G \to G$ given by $g \mapsto g_i g$ (resp. $g \mapsto g g_i$).

PROOF. Definition 3.2.1 is equivalent to the fact that

$$\Delta(\{\varphi, \psi\})(g_1, g_2) = \{\varphi, \psi\}(g_1 g_2) = \langle \pi(g_1 g_2), d\varphi \wedge d\psi \rangle$$

is equal to $\{\Delta(\varphi), \Delta(\psi)\}(g_1, g_2)$. But this is equivalent to (3.2.1), because of formula (1.2.2) for the Poisson structure on $G \times G$. $\square$

COROLLARY 3.2.4. *The unit element of a Poisson Lie group is always a zero-dimensional symplectic leaf.*

PROOF. It follows from the fact that $\pi(e) = 0$. To prove this, it suffices to plug $g_1 = g_2 = e$ into (3.2.1). $\square$

EXERCISE 3.2.5. Let $\eta : G \to \mathfrak{g} \wedge \mathfrak{g}$ be given by

$$(3.2.2) \qquad \eta(g) = \left(r_{g^{-1}}\right)_* \pi(g).$$

Show that (3.2.1) is equivalent to

$$(3.2.3) \qquad \eta(g_1 g_2) = \eta(g_1) + \mathrm{Ad}_{g_1} \eta(g_2).$$

This means that η is a 1-cocycle on G with respect to the adjoint action of G on $\mathfrak{g} \wedge \mathfrak{g}$.

In the following examples the Lie groups are at the same time algebraic groups. In fact, we describe Poisson brackets on the algebras of regular functions. However, it is always clear in those examples how to extend them to the corresponding Poisson brackets to the algebras of C^∞-functions. We leave it as an exercise to the reader.

EXAMPLE 3.2.6. Consider the abelian Lie group $G = \mathbb{C}^n$. It follows immediately from Proposition 3.2.2 that any Poisson Lie group structure is of the form

$$(3.2.4) \qquad \{x_i, x_j\} = \sum_{k=1}^{n} c_{ij}^k x_k,$$

where x_m ($m = 1, \ldots, n$) are the coordinate functions, and the structure constants c_{ij}^k satisfy the following conditions:

$$c_{ij}^k = -c_{ji}^k$$

$$\sum_{l=1}^{n} \left(c_{ij}^l c_{lk}^m + c_{jk}^l c_{li}^m + c_{ki}^l c_{lj}^m\right) = 0.$$

Note that by (1.1.1), it suffices to define the Poisson brackets on the coordinate functions.

Indeed, the definition (1.1.2) of the Poisson bracket in terms of π can be rewritten in terms of η as follows:
$$\{f_1, f_2\} = \sum_{\alpha,\beta} \eta_{\alpha\beta}(g) \partial_\alpha f_1 \partial_\beta f_2,$$
where
$$\eta(g) = \eta^{\alpha\beta}(g) I_\alpha \otimes I_\beta,$$
$\{I_\alpha\}$ being a basis of $\mathfrak{g} = T_e G$ and ∂_α the right-invariant vector field on G which takes the value I_α at e.

When $G = \mathbb{C}^n$, this implies that η is a linear function, hence we have (3.2.4).

EXAMPLE 3.2.7. An important special case of the previous example is as follows. Let $\mathfrak{g}$ be a Lie algebra. Consider the dual vector space $\mathfrak{g}^*$ equipped with the structure of abelian Lie group. Then there exists a canonical Poisson Lie group structure on $\mathfrak{g}^*$ given by the Kirillov–Kostant bracket

(3.2.5) $$\{a, b\} = [a, b],$$

where $a, b \in \mathfrak{g}$ on the right-hand side, and a, b are regarded as linear functions on $\mathfrak{g}^*$ on the left-hand side.

EXAMPLE 3.2.8. It is easy to see that any Lie group G with the trivial Poisson bracket (that is, the Poisson bracket of any two functions is zero) is a Poisson Lie group.

EXAMPLE 3.2.9. Consider the Lie group $G = SL_2(\mathbb{C})$ of complex 2×2 matrices with determinant 1. Show that the following relations define a Poisson Lie group structure on G:
$$\{t_{11}, t_{12}\} = -t_{11}t_{12}, \quad \{t_{11}, t_{21}\} = -t_{11}t_{21},$$
$$\{t_{12}, t_{22}\} = -t_{12}t_{22}, \quad \{t_{21}, t_{22}\} = -t_{21}t_{22},$$
$$\{t_{12}, t_{21}\} = 0, \quad \{t_{11}, t_{22}\} = -2t_{12}t_{21},$$
where t_{ij} ($i, j = 1, 2$) are the matrix elements.

3.3. The correspondence between Poisson Lie groups and Lie bialgebras. One of the most important facts of the Lie theory is the correspondence between Lie groups and Lie algebras. Recall that given a Lie group G, the tangent space at the unit element has a canonical Lie algebra structure. Conversely, given a Lie algebra $\mathfrak{g}$, there exists a unique connected and simply connected Lie group whose tangent space at the unit element is isomorphic to $\mathfrak{g}$ as a Lie algebra.

In this section we establish a Poisson counterpart of this result. Namely, let G be a Poisson Lie group, and $\mathfrak{g}$ its Lie algebra. We can identify $\mathfrak{g}$ with the tangent space $T_e G$ to G at the unit element of the group. Define a linear map $\varphi : \mathfrak{g} \to \mathfrak{g} \wedge \mathfrak{g}$ as the linear map which is dual to the Lie bracket in $\mathfrak{g}^* = T_e^* G$ given by

(3.3.1) $$[\alpha, \beta] = d_e\{f, g\},$$

for any $\alpha, \beta \in \mathfrak{g}^*$, where $f, g \in C^\infty(G)$ are such that
$$d_e f = \alpha, \quad d_e g = \beta.$$

3. POISSON GROUPS

THEOREM 3.3.1. (1) *Let G be a Poisson Lie group. Then there exists a canonical Lie bialgebra structure on the Lie algebra $\mathfrak{g} = T_e G$ with the cobracket $\varphi : \mathfrak{g} \to \mathfrak{g} \wedge \mathfrak{g}$ given by* (3.3.1).

(2) *Let G be a Lie group, and suppose that the Lie algebra $\mathfrak{g} = T_e G$ is equipped with a Lie bialgebra structure. Then there exists a unique Poisson Lie group structure on G such that* (3.3.1) *holds.*

PROOF. The skew-commutativity and the Jacobi identity for the bracket (3.3.1) follow immediately from the corresponding properties of the Poisson bracket $\{,\}$ on $C^\infty(G)$. Thus, we see that it defines a Lie algebra structure on $\mathfrak{g}^*$. We only have to prove that the dual map φ is a 1-cocycle. The cocyclicity of φ follows from (3.2.3). This proves the first statement.

To prove the second statement, it suffices to reconstruct the 1-cocycle $\eta : G \to \mathfrak{g} \wedge \mathfrak{g}$ which defines the Poisson bivector field by (3.2.2). The existence and uniqeness of η follows from the lemma.

LEMMA 3.3.2. *Consider the actions of $\mathfrak{g}$ and G on $\mathfrak{g} \wedge \mathfrak{g}$ given by the coadjoint actions on $\mathfrak{g}$. Then for any 1-cocycle $\varphi : \mathfrak{g} \to \mathfrak{g} \wedge \mathfrak{g}$, there exists a unique 1-cocycle $\eta : G \to \mathfrak{g} \wedge \mathfrak{g}$ such that $\varphi = \frac{d}{dg}|_{g=e} \eta(g)$.*

PROOF. Differentiating the cocyclicity condition

$$\eta(\exp tX \cdot g) = \eta(\exp tX) + \mathrm{Ad}_{\exp tX} \, \eta(g)$$

with respect to t at $t = 0$, we get

$$(3.3.2) \quad \begin{aligned} (\mathcal{L}_{\zeta_X} \eta)(g) &= (\mathcal{L}_{\zeta_X} \eta)(e) + [X \otimes 1 + 1 \otimes X, \eta(g)] \\ &= \varphi(X) + [X \otimes 1 + 1 \otimes X, \eta(g)], \end{aligned}$$

where $\mathcal{L}$ stands for the Lie derivative, and $\zeta : X \to \zeta_X$ is the Lie algebra antihomomorphism from $\mathfrak{g}$ into the Lie algebra of vector fields on G, corresponding to the infinitesimal action of $\mathfrak{g}$ (while G acts by left translations).

Let $\{I_\alpha\}$ be a basis in $\mathfrak{g}$. Plugging $X = I_\alpha$ into (3.3.2), we get a system of differential equations on G with the initial conditions $\eta(e) = 0$.

Consider the vector fields on G with values in $\mathfrak{g} \wedge \mathfrak{g}$ given by

$$(3.3.3) \quad V_\alpha = \mathcal{L}_{\zeta_{I_\alpha}} + [I_\alpha \otimes 1 + 1 \otimes I_\alpha, \cdot].$$

It can be readily checked that the following conditions are satisfied:

- $V_\alpha(e)$ are linearly independent;
- $[V_\alpha, V_\beta] = \sum_\gamma c^\gamma_{\alpha\beta} V_\gamma$ for any α, β, where $c^\gamma_{\alpha\beta}$ are numbers depending on the point in a neighborhood of the unit element.

By the Frobenius theorem applied to the vector fields on G with coefficients in $\mathfrak{g} \wedge \mathfrak{g}$, we get that there exists a fundamental solution for the system of partial differential equations defined by the vector fields (3.3.3) in a neighborhood of the unit element in G. On the other hand, since I_α form a basis in $\mathfrak{g}$, the above fundamental solution is unique.

It follows that there exists a unique solution to (3.2.3) in a neighborhood of the unit element. Since G is connected, η can be extended to the whole group by applying (3.2.3). □

We leave it to the reader to prove the Jacobi identity. □

PROPOSITION 3.3.3. (1) *The correspondence $G \mapsto \mathfrak{g} = T_e G$ establishes a covariant functor from the category of Poisson Lie groups to the category of Lie bialgebras.*

(2) *This correspondence establishes an equivalence between the full subcategory of connected and simply connected Poisson Lie groups and the category of Lie bialgebras.*

PROOF. For any Poisson Lie group G, define the Lie bialgebra structure on $\mathfrak{g}$ as in Theorem 3.3.1. Any Poisson Lie group homomorphism $G_1 \to G_2$ induces a homomorphism $C^\infty(G_2) \to C^\infty(G_1)$ of Poisson Hopf algebras. At the same time, it defines the map of the tangent spaces $\mathfrak{g}_1 \to \mathfrak{g}_2$. The fact that it is a Lie algebra homomorphism follows from (3.3.1) and the Leibniz rule (1.1.1). The first statement is proved.

The above covariant functor can be obviously restricted to the full subcategory of connected and simply connected Poisson Lie groups. On the other hand, for any Lie algebra $\mathfrak{g}$, there exists a unique connected and simply connected Lie group G which has $\mathfrak{g}$ as its Lie algebra. By Theorem 3.3.1, there is a unique Poisson Lie group structure on G such that $\mathfrak{g}$ is the corresponding Lie bialgebra. We leave it as an exercise to the reader to check that this is the inverse functor. □

EXAMPLE 3.3.4. (1) The Lie bialgebra of Example 2.1.6 is the Lie bialgebra of the Poisson Lie group of Example 3.2.8.

(2) The Lie bialgebra of Example 2.1.7 is the Lie bialgebra of the Poisson Lie group of Example 3.2.7.

(3) The Lie bialgebra of Example 2.1.8 is the Lie bialgebra of the Poisson Lie group of Example 3.2.9.

DEFINITION 3.3.5. Let G be a Poisson Lie group, $\mathfrak{g}$ the Lie bialgebra of G.

(1) The connected and simply connected Poisson Lie group G^* which corresponds to the dual Lie bialgebra $\mathfrak{g}^*$ is called *dual Poisson Lie group*.

(2) The connected and simply connected Poisson Lie group $\mathcal{D}(G)$ which corresponds to the double Lie bialgebra $\mathcal{D}(\mathfrak{g})$ is called *double Poisson Lie group*.

EXERCISE 3.3.6. Suppose that G is the Poisson Lie group with the trivial (zero) Poisson bracket (cf. Example 3.2.8). Then the dual Poisson Lie group is isomorphic to $\mathfrak{g}^*$ regarded as an abelian Lie group and equipped with the Kirillov–Kostant bracket (3.2.5) (cf. Example 3.2.7).

EXAMPLE 3.3.7. By Theorem 3.3.1, a standard Lie bialgebra structure on a real form $\mathfrak{g}_0$ of a complex simple Lie algebra $\mathfrak{g}$ (cf. Exercise 2.3.9) induces a Poisson Lie group structure on the real form G_0 of the complex simple Lie group G such that the Lie algebra of G_0 is isomorphic to $\mathfrak{g}_0$ and the Lie algebra of G to $\mathfrak{g}$. (Recall that a *real form* of G is a real Lie subgroup $G_0 \subset G$ such that G is isomorphic to the complexification of G_0.)

EXAMPLE 3.3.8. When $G_0 = K$ is a compact real form of G, a standard Poisson Lie group structure on G_0 is unique up to conjugation. Consider the Iwasawa decomposition $G = KAN_+$, where $A = \exp(it)$. By Exercise 2.3.9, the dual Poisson Lie group K^* is isomorphic to $AN_+ \subset G$, and the multiplication maps

$$K \times K^* = K \times AN_+ \to G,$$
$$K^* \times K = AN_+ \times K \to G$$

are Poisson diffeomorphisms.

3. POISSON GROUPS

DEFINITION 3.3.9. The Poisson Lie group structure on G_0 described in Example 3.3.7 is called *standard*.

EXAMPLE 3.3.10. Consider the Poisson Lie group $SU(p,q)$ equipped with a standard Poisson Lie group structure (cf. Exercise 2.3.9). As follows from Exercise 2.3.11, the dual Poisson Lie group is isomorphic as a Lie group to the group of complex upper-triangular matrices with positive diagonal entries and determinant 1.

3.4. Symplectic leaves in Poisson Lie groups and dressing action.

Now we show how to use Theorem 1.2.5 to describe symplectic leaves in a Poisson Lie group. The double Poisson Lie group construction allows us to describe the symplectic leaves locally as the orbits of the so-called dressing action. Throughout the section the word "locally" means "in a neighborhood of the unit element of the group."

Let G be a Poisson Lie group, and $\mathfrak{g}$ the Lie bialgebra of G. Recall that Proposition 2.3.6 gives a simple description of the dual Lie bialgebra structure on $\mathcal{D}(\mathfrak{g})$ by (2.3.3) where $r = \sum_\alpha I_\alpha \otimes I^\alpha$, $\{I_\alpha\}$ being a basis of $\mathfrak{g}$ and $\{I^\alpha\}$ the dual basis of $\mathfrak{g}^*$. The following proposition is a special case of a more general result (see Proposition 4.1.4). Thus, we omit the proof here.

PROPOSITION 3.4.1. *The Poisson bracket on* $\mathcal{D}(G)$ *is given by*
$$(3.4.1)\quad \{f_1, f_2\} = \sum_\alpha \left(\partial_\alpha f_1 \partial^\alpha f_2 - \partial^\alpha f_1 \partial_\alpha f_2\right) - \sum_\alpha \left(\partial'_\alpha f_1 (\partial^\alpha)' f_2 - (\partial^\alpha)' f_1 \partial'_\alpha f_2\right),$$

where ∂_α (resp. ∂'_α) is the right (resp. left) invariant vector field on $\mathcal{D}(\mathfrak{g})$ which takes the value I_α at the unit element of $\mathcal{D}(G)$, while ∂^α (resp. $(\partial^\alpha)'$) is the right (resp. left) invariant vector field on $\mathcal{D}(G)$ which takes the value I^α at the unit element of $\mathcal{D}(G)$.

PROPOSITION 3.4.2. *The multiplication maps*
$$(3.4.2) \quad m: G \times G^* \to \mathcal{D}(G),$$
$$(3.4.3) \quad m: G^* \times G \to \mathcal{D}(G)$$

are local Poisson diffeomorphisms in a neighborhood of the unit element.

PROOF. This is an easy consequence of the fact that $\mathcal{D}(\mathfrak{g}) = \mathfrak{g} \oplus \mathfrak{g}^*$ as a vector space. $\square$

It is clear that one can identify locally G with $G^* \backslash \mathcal{D}(G)$ or with $\mathcal{D}(G)/G^*$. Consider the natural projections $p_1 : \mathcal{D}(G) \to G^* \backslash \mathcal{D}(G)$ and $p_2 : \mathcal{D}(G) \to \mathcal{D}(G)/G^*$. Then we have the following property.

PROPOSITION 3.4.3. *The Poisson manifold structure on the double Lie group $\mathcal{D}(G)$ induces Poisson manifold structures on $G^* \backslash \mathcal{D}(G)$ and $\mathcal{D}(G)/G^*$ such that the natural projections $p_1 : \mathcal{D}(G) \to G^* \backslash \mathcal{D}(G)$ and $p_2 : \mathcal{D}(G) \to \mathcal{D}(G)/G^*$ are Poisson maps. Both manifolds are isomorphic to G as Poisson manifolds in a neighborhood of the coset G^*.*

PROOF. By Proposition 3.4.2,
$$\{f_1, f_2\}_{\mathcal{D}(G)}(gh) = \{f_1(gh), f_2(gh)\}_G + \{f_1(gh), f_2(gh)\}_{G^*}$$

for any $g \in G$ and $h \in G^*$. In particular, for any right G^*-invariant functions f_1 and f_2, we have
$$\{f_1, f_2\}_{\mathcal{D}(G)} = \{f_1, f_2\}_G.$$
This means that if we define a Poisson manifold structure on $\mathcal{D}(G)$ by the bivector field
$$\pi_{\mathcal{D}(G)/G^*} = (p_2)_* \pi_{\mathcal{D}(G)},$$
it is equivalent to the Poisson manifold structure on G with respect to the identification $G \simeq \mathcal{D}/G^*$ in a neighborhood of the unit element.

The case of $G^* \backslash \mathcal{D}(G)$ can be considered similarly. □

Now we introduce the notion of left and right dressing actions. First we define them locally, in a neighborhood of the unit element. As follows from Proposition 3.4.2, for any given $g \in G$ and $h \in G^*$ which lie in some neighborhoods of the unit element of $\mathcal{D}(G)$, there exist unique $g^h \in G$ and $h^g \in G^*$ such that

(3.4.4) $$hg = g^h h^g.$$

This formula defines a local left action of G^* on G and a local right action of G on G^* given by

(3.4.5) $$h : g \mapsto g^h, \quad g : h \mapsto h^g,$$

respectively. Note that we can replace G by G^* and vice versa, so that we have also a local right action of G^* on G and a local left action of G on G^*.

DEFINITION 3.4.4. The local left (resp. right) action of G on G^* defined by (3.4.4), (3.4.5) is called *local left* (resp. *right*) *dressing action* of G on G^*. If it can be extended to a global action, the latter is called *global left* (resp. *right*) *dressing action* of G on G^*.

Now we will give another description of the left and right dressing actions in terms of the dressing fields. So far they have been defined as local actions of G^* on G in a neighborhood of unity in G. The new definition allows us to extend them to local actions in a neighborhood of any element g in G, even though still in a neighborhood (depending on g, of course) of the unit element in G^*.

Indeed, to talk about a local action of a group G on a manifold is the same as to consider a representation of the Lie algebra $\mathfrak{g}$ of G by vector fields on the manifold. Recall also that as a vector space $\mathfrak{g}^*$ is isomorphic to the space of left (or right) invariant 1-forms on G, the correspondence being given by

$$\alpha \in \mathfrak{g}^* \mapsto \alpha_l, \quad \alpha \in \mathfrak{g}^* \mapsto \alpha_r,$$

where
$$\alpha_l(e) = \alpha_r(e) = \alpha \in T_e^* G \simeq \mathfrak{g}^*.$$

EXERCISE 3.4.5. Let G be a Poisson Lie group, π the corresponding bivector field. The left (or right) invariant 1-forms on G form a Lie subalgebra of $\Omega^1(G)$ with respect to the Lie bracket (1.2.3).

The corresponding Lie algebra structure on $\mathfrak{g}^*$ coincides with the one given by linearizing π at the unit element of G.

Consider the linear maps $\rho_l, \rho_r : \mathfrak{g}^* \to Vect(G)$ given by

(3.4.6) $$\rho_l(\alpha) = \check{\pi} \alpha_l, \quad \rho_r(\alpha) = -\check{\pi} \alpha_r.$$

By Propositions 1.2.14 and 3.4.5, ρ_l is a Lie algebra antihomomorphism, while ρ_r is a Lie algebra homomorphism.

DEFINITION 3.4.6. We call $\rho_l(\alpha)$ (resp. $\rho_r(\alpha)$) *left* (resp. *right*) *dressing field on G corresponding to $\alpha \in \mathfrak{g}^*$*.

PROPOSITION 3.4.7. *Consider the infinitesimal action $\alpha \mapsto \check{\rho}_l(\alpha)$ (resp. $\alpha \mapsto \check{\rho}_r(\alpha)$) of $\mathfrak{g}^*$ by vector fields in a neighborhood of the unit element in G which corresponds to the left (resp. right) dressing action of G^* on G as defined in Definition 3.4.4. Then*

$$\check{\rho}_l(\alpha) = \rho_l(\alpha), \quad \check{\rho}_r(\alpha) = \rho_r(\alpha)$$

in a neighborhood of the unit element in G for any $\alpha \in \mathfrak{g}^*$.

PROOF. In order to prove this proposition, we will show that the dressing vector fields satisfy certain properties that characterize them uniquely. Then we show that the vector fields that correspond to the dressing actions satisfy the same properties.

LEMMA 3.4.8. *A Lie algebra homomorphism (resp. antihomomorphism) τ_r (resp. τ_l) from $\mathfrak{g}^*$ to the Lie algebra of vector fields on G coincides with the right (resp. left) dressing fields $\rho_r : \alpha \mapsto \rho_r(\alpha)$ (resp. $\rho_l : \alpha \mapsto \rho_l(\alpha)$) if and only if it satisfies the following properties.*

(1) *For any $\alpha \in \mathfrak{g}^*$ and $g, h \in G$, we have*

$$\begin{aligned}\tau_l(\alpha)(gh) &= (l_g)_* \tau_l(\alpha)(h) + (r_h)_* \tau_l\left(\mathrm{Ad}^*_{h^{-1}} \alpha\right)(g), \\ \tau_r(\alpha)(gh) &= (l_g)_* \tau_r\left(\mathrm{Ad}^*_g \alpha\right)(h) + (r_h)_* \tau_r(\alpha)(g).\end{aligned}$$

(2) *For any $\alpha \in \mathfrak{g}^*$, consider the linearization of $\check{\rho}_l(\alpha)$ (resp. of $\check{\rho}_r(\alpha)$) at the unit element of G given by the following linear maps on $\mathfrak{g}$:*

(3.4.7) $$\xi \mapsto \left[\tilde{\xi}, \check{\rho}_l(\alpha)\right](e),$$

(3.4.8) $$\xi \mapsto \left[\tilde{\xi}, \check{\rho}_r(\alpha)\right](e),$$

where $\tilde{\xi}$ is any vector field on G with $\tilde{\xi}(e) = \xi$.

Then the linearization (3.4.7) of $\check{\rho}_l(\alpha)$ defines the coadjoint representation of $\mathfrak{g}^*$ in $\mathfrak{g}$, while the linearization (3.4.8) of $\check{\rho}_r(\alpha)$ defines the antirepresentation of $\mathfrak{g}^*$ on $\mathfrak{g}$ which is minus the coadjoint representation.

PROOF. First we prove that the left and right dressing fields $\rho_l(\alpha)$ and $\rho_r(\alpha)$ satisfy properties (1) and (2). Property (1) is a direct consequence of (3.2.1). Let us show that (2) holds.

Since $\rho_l(\alpha)$ vanishes at the unit element e of G for any $\alpha \in \mathfrak{g}^*$, it is linearized at e to give a map

$$\mathfrak{g} \to \mathfrak{g}, \quad X \mapsto [X_r, \rho_l(\alpha)](e),$$

where X_r is the right-invariant vector field on G such that $X_r(e) = X \in \mathfrak{g}$. For any $\beta \in \mathfrak{g}^*$ we have

$$\begin{aligned}[X_r, \rho_l(\alpha)](e)(\beta) &= \frac{d}{dt}\Big|_{t=0} \left(l_{\exp(-tX)}\right)_* \rho_l(\alpha) e^{tX}(\beta) \\ &= \frac{d}{dt}\Big|_{t=0} \pi_l\left(e^{tX}\right)(\beta, \alpha) = X\left([\beta, \alpha]\right) = (\mathrm{ad}^*_\alpha X)(\beta).\end{aligned}$$

This shows that $[X_r, \rho_l(\alpha)](e) = \mathrm{ad}^*_\alpha X$. The case of the right dressing fields can be considered similarly.

Finally, suppose that there is another Lie algebra antihomomorphism ρ from $\mathfrak{g}^*$ into the Lie algebra of vector fields on G which satisfies the properties (1) and (2). Consider the difference $\rho_0 = \rho - \rho_l$. It obviously satisfies (1). For any $\alpha, \beta \in \mathfrak{g}^*$, consider the corresponding left invariant 1-forms α_l and β_l. The bivector field π_0 on G given by
$$\pi_0(\alpha_l, \beta_l) = \langle \alpha_l, \rho_0(\beta) \rangle$$
can be easily checked to satisfy (3.2.1). Furthermore, since $[\xi, \rho_0(\alpha)](e) = 0$ for any vector field ξ on G and $\alpha \in \mathfrak{g}^*$, the linearization of π_0 at e is equal to zero. We leave it as a simple exercise to deduce that $\pi_0 = 0$. □

By (2.3.1), the linearization at the unit element of G of the left (resp. right) dressing action of G^* on G is the coadjoint (resp. minus coadjoint) representation of $\mathfrak{g}^*$ on $\mathfrak{g}$. Thus, we only have to check that the corresponding vector fields satisfy the property (1). But it can be easily shown by differentiating the identity
$$\exp(s\xi_1)\exp(t\xi_2) = (\exp(t\xi_2))^{\exp(s\xi_1)} (\exp(s\xi_1))^{\exp(t\xi_2)}$$
with respect to $t = 0$ and $s = 0$ successively, where $\xi_1 \in \mathfrak{g}^*$ and $\xi_2 \in \mathfrak{g}$. We leave the details to the reader. □

Note that Definition 3.4.4 defines the left and right dressing actions only of a small neighborhood of the unit element of G^* in a small neighborhood of the unit element of G. Proposition 3.4.7 allows us to redefine those actions for a small neighborhood of any element in G.

DEFINITION 3.4.9. The local action of G^* in a neighborhood of an element g in G given by the left (resp. right) dressing fields $\rho_l(\alpha)$ (resp. $\rho_r(\alpha)$), where $\alpha \in \mathfrak{g}^*$, is called the *left* (resp. *right*) *dressing action* of G^* on G.

PROPOSITION 3.4.10. *Let G be a Poisson Lie group and G^* the Poisson Lie group dual to G. Then the symplectic leaves in G locally coincide with the orbits of the right (or left) dressing action of G^*. In particular, if the right (resp. left) dressing action can be defined globally, the symplectic leaves in G are the orbits of the right (resp. left) dressing action.*

PROOF. Indeed, by Proposition 3.4.7, the left dressing action can be defined in terms of the left dressing vector fields. By definition, they are tangent to the symplectic leaves. On the other hand, it is clear that the values of $\rho_l(\alpha)$ at any $g \in G$ span the tangent space $T_g S = \tilde{\pi}\left(T_g^* G\right)$ to the symplectic leaf S passing through g. It follows that the symplectic leaves locally coincide with the orbits of the dressing action.

If the left dressing action can be defined globally, we conclude that for any orbit there is a symplectic leaf in G that contains it. On the other hand, any submanifold of a Poisson manifold which has the tangent space at each point isomorphic to the image of $\tilde{\pi}$ is a Poisson submanifold. Therefore, each orbit of the left dressing action is a Poisson submanifold. Since it is contained in a symplectic leaf, it coincides with it.

The case of the right dressing action can be considered similarly. □

EXERCISE 3.4.11. Suppose that K is a compact Poisson Lie group. Show that there exists the global left (resp. right) dressing action of K on the dual Poisson Lie group K^*. Similarly, there exists the global left (resp. right) dressing action of K^* on K.

Hint: This follows from the Iwasawa decomposition. See also Remark 3.4.12.

REMARK 3.4.12. Let K be a compact real form of a complex semisimple Lie group G equipped with the standard Poisson Lie group structure (cf. Example 3.3.7). Then the dual Lie group K^* is isomorphic to the subgroup $AN_+ \subset G$ where $A = \exp\mathfrak{a}$, $N_+ = \exp\mathfrak{n}_+$, and $\mathfrak{a} = i(\mathfrak{h} \cap \mathfrak{k})$, where $\mathfrak{k}$ is the Lie algebra of K. (For the notation, see Exercise 2.3.9.)

We see that in this case the Iwasawa decomposition

$$G = KAN_+$$

is a global version of (3.4.2)–(3.4.3). Therefore, the dressing actions can be defined globally.

EXERCISE 3.4.13. Suppose that G is the Poisson Lie group with the trivial (zero) Poisson bracket. As we know from Example 3.3.6, the dual Poisson Lie group is isomorphic to $G^* \simeq \mathfrak{g}^*$ regarded as an abelian Lie group equipped with the Kirillov–Kostant bracket (3.2.5).

Show that in this case the left dressing action (resp. the right dressing action transformed to a left action) of G on G^* is defined globally and coincides with the coadjoint action of G on $\mathfrak{g}^*$. In particular, we can see that the symplectic leaves in $\mathfrak{g}^*$ are the coadjoint orbits.

THEOREM 3.4.14. *Let $g \in G$ belong to a sufficiently small neighborhood of the unit element in G. The symplectic leaf in G passing through g locally (in some neighborhood of g) is the image of the double coset $G^*gG^* \subset \mathcal{D}(G)$ under the natural projection $\mathcal{D}(G) \to \mathcal{D}(G)/G^* \simeq G$.*

PROOF. By Proposition 3.4.10, the symplectic leaves coincide locally with the orbits of the left dressing action of G^* on G. (Note that it was deduced from the second definition of the dressing action.) By Definition 3.4.4 of the dressing action (3.4.5), they are the images G^*gG^*/G^* of the double cosets in $\mathcal{D}(G)$ under the natural projection $\mathcal{D}(G) \to \mathcal{D}(G)/G^* \simeq G$. □

4. Lie bialgebras and classical r-matrices

In the present section we study certain special classes of Lie bialgebras. We are particularly interested in quasi-triangular ones. The Belavin–Drinfeld theorem gives a partial classification of them.

4.1. Coboundary, quasi-triangular and triangular Lie bialgebras.
Recall that the central point in the definition of Lie bialgebra is the compatibility condition between the Lie algebras $\mathfrak{g}$ and $\mathfrak{g}^*$, which is expressed by saying that the cobracket $\varphi : \mathfrak{g} \to \mathfrak{g} \wedge \mathfrak{g}$ is a 1-cocycle on $\mathfrak{g}$ (cf. Definition 2.1.1). If we require φ to be a coboundary, this determines a special class of Lie bialgebras.

PROPOSITION 4.1.1. *Let $\mathfrak{g}$ be a Lie algebra, and $r \in \mathfrak{g} \wedge \mathfrak{g}$. Consider the linear map $\varphi : \mathfrak{g} \to \mathfrak{g} \wedge \mathfrak{g}$ given by*

(4.1.1) $$\varphi(a) = [a \otimes 1 + 1 \otimes a, r],$$

for any $a \in \mathfrak{g}$. Then $(\mathfrak{g}, \varphi)$ is a Lie bialgebra if and only if the tensor

(4.1.2) $$\langle r, r \rangle \stackrel{\mathrm{def}}{=} \left[r^{12}, r^{13}\right] + \left[r^{13}, r^{23}\right] + \left[r^{12}, r^{23}\right] \in \mathfrak{g} \otimes \mathfrak{g} \otimes \mathfrak{g}$$

is ad-*invariant*. Here we use the usual notation: $r^{12} = r \otimes 1$, $r^{23} = 1 \otimes r$, and $r^{13} = \sum_k a_k \otimes 1 \otimes b_k$ if $r = \sum_k a_k \otimes b_k$. The commutators on the right-hand side of (4.1.2) are taken in $(U\mathfrak{g})^{\otimes 3}$.

PROOF. The cocyclicity of φ follows immediately from the Jacobi identity for $\mathfrak{g}$. Thus, we only have to prove that the dual linear map $\varphi^* : \alpha \otimes \beta \mapsto [\alpha, \beta]$, where $\alpha, \beta \in \mathfrak{g}^*$ satisfies the Jacobi identity for $\mathfrak{g}^*$ if and only if $\langle r, r \rangle$ is ad-invariant.

Let $\langle , \rangle : \mathfrak{g} \otimes \mathfrak{g}^* \to \mathbf{k}$ be the natural pairing between $\mathfrak{g}$ and $\mathfrak{g}^*$. This is a straightforward computation to check that

$$\langle a, [\alpha, [\beta, \gamma]] + [\beta, [\gamma, \alpha]] + [\gamma, [\alpha, \beta]] \rangle = \langle \mathrm{ad}_a \langle r, r \rangle, \alpha \otimes \beta \otimes \gamma \rangle,$$

which proves the proposition. □

REMARK 4.1.2. Formula (4.1.1) means that

$$\varphi = \partial r,$$

in particular, that φ is a coboundary.

Note also that if we consider elements of $\mathfrak{g} \wedge \mathfrak{g}$ as left (or right) invariant bivector fields on the corresponding Lie group G, then $\langle r, r \rangle$ is nothing but the Schouten bracket of r with itself.

DEFINITION 4.1.3. (1) Equation (4.1.2) is called the *modified classical Yang–Baxter equation* (MCYBE), and a solution to (4.1.2) is called the *classical r-matrix*.

(2) The equation

(4.1.3) $$\langle r, r \rangle = 0$$

is called the *classical Yang–Baxter equation* (CYBE).

Given a solution to (4.1.3), one can construct explicitly the Poisson bracket on a Lie group G with the Lie algebra $\mathfrak{g}$. Choose a basis $\{I_\alpha\}$ in $\mathfrak{g}$, and let ∂_α (resp. ∂'_α) be the right (resp. left) invariant vector field on G whose value at the unit element is I_α. Let $r = \sum_{\alpha, \beta} r_{\alpha\beta} I_\alpha \otimes I_\beta$.

PROPOSITION 4.1.4. *The formula*

(4.1.4) $$\{f_1, f_2\} = \sum_{\alpha, \beta} r_{\alpha\beta} \left(\partial_\alpha f_1 \partial_\beta f_2 - \partial'_\alpha f_1 \partial'_\beta f_2 \right)$$

defines a Poisson–Lie group structure on G such that the canonical Lie bialgebra structure on $\mathfrak{g}$ described in Theorem 3.3.1 coincides with the one defined in terms of r.

PROOF. The skew-commutativity of the bracket (4.1.4), as well as the Leibniz formulas follow directly from the definitions. To prove the Jacobi identity, consider the left and right brackets

$$\{f_1, f_2\}_l = \sum_{\alpha, \beta} r_{\alpha\beta} \partial_\alpha f_1 \partial_\beta f_2,$$

$$\{f_1, f_2\}_r = \sum_{\alpha, \beta} r_{\alpha\beta} \partial'_\alpha f_1 \partial'_\beta f_2.$$

It follows from (4.1.3) that both the left and the right brackets satisfy the Jacobi identity. But so do their linear combinations, since the left and right vector fields commute with each other. This proves that (4.1.4) defines a Poisson manifold structure on G.

The statement that the bracket (4.1.4) is multiplicative (that is, defines a Poisson Lie group structure) on G follows from the fact that the bracket is defined as the difference of left and right invariant parts. Indeed, we have that

$$\{f_1(xy), f_2(xy)\}_{G\times G} = \{f_1(xy), f_2(xy)\}_x + \{f_1(xy), f_2(xy)\}_y$$

$$= \sum_{\alpha,\beta} r_{\alpha\beta} \left(\partial_{\alpha,x} f_1(xy) \partial_{\beta,x} f_2(xy) - \partial'_{\alpha,x} f_1(xy) \partial'_{\beta,x} f_2(xy)\right)$$

$$+ \sum_{\alpha,\beta} r_{\alpha\beta} \left(\partial_{\alpha,y} f_1(xy) \partial_{\beta,y} f_2(xy) - \partial'_{\alpha,y} f_1(xy) \partial'_{\beta,y} f_2(xy)\right)$$

$$= \sum_{\alpha,\beta} r_{\alpha\beta} \left(\partial_{\alpha,x} f_1(xy) \partial_{\beta,x} f_2(xy) - \partial'_{\alpha,y} f_1(xy) \partial'_{\beta,y} f_2(xy)\right)$$

$$+ \sum_{\alpha,\beta} r_{\alpha\beta} \left(\partial_{\alpha,y} f_1(xy) \partial_{\beta,y} f_2(xy) - \partial'_{\alpha,x} f_1(xy) \partial'_{\beta,x} f_2(xy)\right)$$

$$= \sum_{\alpha,\beta} r_{\alpha\beta} \left(\partial_\alpha f_1 \partial_\beta f_2 - \partial'_\alpha f_1 \partial'_\beta f_2\right)(xy) = \{f_1, f_2\}_G(xy),$$

where $\partial_{\alpha,x}$, $\partial_{\alpha,y}$, $\partial'_{\alpha,x}$, and $\partial'_{\alpha,y}$ mean the left and right differentiations in x and y respectively. $\square$

EXERCISE 4.1.5. Let $\mathfrak{g}$ be a Lie algebra, and $r \in \mathfrak{g} \otimes \mathfrak{g}$. Suppose that r satisfies CYBE (4.1.3) and

(4.1.5) $$u = r^{12} + r^{21}$$

is ad-invariant. Show that $\varphi = \partial r$ maps $\mathfrak{g}$ to $\mathfrak{g} \wedge \mathfrak{g}$ and $(\mathfrak{g}, \varphi)$ is a Lie bialgebra.

EXERCISE 4.1.6. (1) Show that if r satisfies (4.1.3) and (4.1.5), then putting $\tilde{r} = r - \frac{1}{2}u$, we get that $\tilde{r} \in \mathfrak{g} \wedge \mathfrak{g}$, $\varphi = \partial \tilde{r}$, and

$$\langle \tilde{r}, \tilde{r} \rangle = \frac{1}{4}\left[u^{12}, u^{23}\right].$$

(2) Show that if $\mathfrak{g}$ is a complex simple Lie algebra, then $r \in \mathfrak{g} \otimes \mathfrak{g}$ satisfies (4.1.3) and (4.1.5) if and only if $\tilde{r} = r - \frac{1}{2}u$ satisfies MCYBE (4.1.2). Hint: the subspace of ad-invariant elements in $\mathfrak{g} \wedge \mathfrak{g} \wedge \mathfrak{g}$ is one-dimensional.

DEFINITION 4.1.7. (1) A Lie bialgebra $(\mathfrak{g}, \varphi)$ is called a *coboundary Lie bialgebra* if there exists $r \in \mathfrak{g} \wedge \mathfrak{g}$ such that $\varphi = \partial r$.

(2) A coboundary Lie bialgebra $(\mathfrak{g}, \varphi)$ is called a *triangular Lie bialgebra* if r satisfies CYBE (4.1.3).

(3) A Lie bialgebra $(\mathfrak{g}, \varphi)$ is called a *quasi-triangular Lie bialgebra* if there exists $r \in \mathfrak{g} \otimes \mathfrak{g}$ which satisfies CYBE (4.1.3) and $\varphi = \partial r$.

(4) In all the above cases r is called the *classical r-matrix*.

REMARK 4.1.8. Note that if $(\mathfrak{g}, \varphi, r)$ is a quasi-triangular Lie bialgebra, then r satisfies (4.1.5). In particular, as follows from Exercise 4.1.6, any quasi-triangular Lie bialgebra is coboundary.

EXAMPLE 4.1.9. By Proposition 2.3.6, for any Lie bialgebra $\mathfrak{g}$, the double Lie bialgebra $\mathcal{D}(\mathfrak{g})$ is always quasi-triangular. Moreover, the canonical element of $\mathfrak{g} \otimes \mathfrak{g}^* \subset \mathcal{D}(\mathfrak{g}) \otimes \mathcal{D}(\mathfrak{g})$ is a classical r-matrix.

EXAMPLE 4.1.10. The Lie bialgebra $\mathfrak{g} = \mathfrak{sl}(2,\mathbb{C})$ introduced in Example 2.1.8 is a quasi-triangular Lie bialgebra, with the classical r-matrix given by

$$r = X^+ \otimes X^- + \frac{1}{4} H \otimes H \tag{4.1.6}$$

(in this case $\tilde{r} = \frac{1}{2} X^+ \wedge X^-$).

EXERCISE 4.1.11. Let $\mathfrak{g}$ be a complex simple Lie algebra equipped with the standard Lie bialgebra structure (cf. Example 2.1.9). Show that it is quasi-triangular, with the classical r-matrix given by

$$r = \sum_{\alpha > 0} X_\alpha \wedge X_{-\alpha} \tag{4.1.7}$$

where X_α is the vector from the root subspace coresponding to the root $\alpha \in \mathfrak{g}^*$, chosen so that $(X_\alpha, X_{-\alpha}) = 1$ for $\alpha > 0$ (where $(\ ,\ )$ is a fixed nondegenerate invariant symmetric bilinear form on $\mathfrak{g}$).

4.2. The classification of quasi-triangular Lie bialgebras. Before we formulate the Belavin–Drinfeld theorem on the classification of quasi-triangular complex simple Lie bialgebras, let us consider a much simpler problem of the classification in dimension 2.

As is well known, there are only two 2-dimensional Lie algebras up to isomorphism—the abelian one, denoted by $\mathfrak{a}$, and the Lie algebra $\mathfrak{b}$ generated by h and x with the relation $[h, x] = x$.

PROPOSITION 4.2.1. *The following is a full list of equivalence classes of two-dimensional Lie bialgebras:*
(1) $(\mathfrak{a}, 0)$ *(the trivial cobracket)*;
(2) $(\mathfrak{b}, 0)$ *(the trivial cobracket)*;
(3) *the Lie bialgebra* $(\mathfrak{a}, \varphi_\mathfrak{a})$ *dual to* $(\mathfrak{b}, 0)$;
(4) $(\mathfrak{b}, \varphi_\mathfrak{b})$, *where* $\varphi_\mathfrak{b}(x) = 0$ *and* $\varphi_\mathfrak{b}(h) = x \wedge h$;
(5) $(\mathfrak{b}, \varphi'_\mathfrak{b})$, *where* $\varphi'_\mathfrak{b}(h) = 0$ *and* $\varphi'_\mathfrak{b}(x) = x \wedge h$.

PROOF. First, consider the abelian Lie algebra $\mathfrak{a}$. If the cobracket δ is nontrivial, it is a linear map from the two-dimensional vector space $\mathfrak{a}$ into the one-dimensional vector space $\mathfrak{a} \wedge \mathfrak{a}$. Therefore, it should have a one-dimensional kernel. It is easy to see that one can always choose a basis x, h in $\mathfrak{a}$ such that $\varphi(h) = 0$ and $\varphi(x) = h \wedge x$. This cobracket obviously defines a Lie bialgebra, which is dual to the Lie bialgebra $(\mathfrak{b}, 0)$.

Consider now the Lie algebra $\mathfrak{b}$. If the cobracket is nontrivial, there are two possibilities: $\varphi(x) = 0$ and $\varphi(x) \neq 0$. In the first case, without losing generality (that is, up to a suitable change of the basis), we may assume that $\varphi(h) = x \wedge h$, and in the second case we may assume that $\varphi(h) = 0$ and $\varphi(x) = x \wedge h$. It is easy to check that both cobrackets define the Lie bialgebra structures. Moreover, it is easy to see that the first one is given by the classical r-matrix $r = x \wedge h$. □

The Belavin–Drinfeld theorem deals with the case when $\mathfrak{g}$ is a finite-dimensional complex simple Lie algebra. Recall that by Exercise 4.1.6 in this case the quasi-triangularity conditions (4.1.3), (4.1.5) on $r \in \mathfrak{g} \otimes \mathfrak{g}$ are equivalent to the condition (4.1.2) on $\tilde{r} \in \mathfrak{g} \wedge \mathfrak{g}$.

The goal is to describe all equivalence classes of quasi-triangular Lie bialgebra structures on $\mathfrak{g}$. This problem has not been solved in full. It ramifies in two principal cases:

(1) $r^{12} + r^{21} \neq 0$;
(2) $r^{12} + r^{21} = 0$.

The theorem provides the answer in the first case only. Since all ad-invariant elements of $\mathfrak{g} \otimes \mathfrak{g}$ are proportional to the canonical element $t = \sum_k I_k \otimes I_k$ where $\{I_k\}$ is an orthonormal basis of $\mathfrak{g}$ with respect to a nondegenerate invariant symmetric bilinear form, the problem reduces to solving the following system of equations on $r \in \mathfrak{g} \otimes \mathfrak{g}$:

(4.2.1) $$r^{12} + r^{21} = t,$$
(4.2.2) $$\left[r^{12}, r^{13}\right] + \left[r^{12}, r^{23}\right] + \left[r^{13}, r^{23}\right] = 0.$$

The second (triangular) case is much more difficult. It contains, for instance, a subproblem of classification of all so-called Frobenius Lie subalgebras (see Subsection 4.3).

SETTING OF THE BELAVIN–DRINFELD THEOREM. Let $\mathfrak{g}$ be a complex simple Lie algebra with a fixed nondegenerate invariant symmetric bilinear form $(\,,\,)$. Fix a Cartan subalgebra $\mathfrak{h}$ and a set of simple roots $\Gamma \subset \mathfrak{h}^*$. Let $t \in \mathfrak{g} \otimes \mathfrak{g}$ be the canonical element with respect to $(\,,\,)$, and $t_0 \in \mathfrak{h} \otimes \mathfrak{h}$ the "$\mathfrak{h}$-component" of t, i.e. the canonical element with respect to the restriction of $(\,,\,)$ to $\mathfrak{h}$.

The solutions to the system (4.2.1), (4.2.2) depend on certain discrete and continuous parameters. The discrete one is a triple $(\Gamma_1, \Gamma_2, \tau)$, where $\Gamma_1, \Gamma_2 \in \Gamma$ and $\tau : \Gamma_1 \to \Gamma_2$ is a one-to-one map such that the following conditions are satisfied:
(1) $(\tau(\alpha), \tau(\beta)) = (\alpha, \beta)$ for any $\alpha, \beta \in \Gamma_1$;
(2) for any $\alpha \in \Gamma_1$, there exists $k \in \mathbb{N}$ such that $\tau^k(\alpha) \notin \Gamma_1$.

Note that $\tau^k(\alpha)$ is defined only if $\alpha, \tau(\alpha), \ldots, \tau^{k-1}(\alpha) \in \Gamma_1$. Thus, the second condition on τ means that τ^k is not defined for a sufficiently large k. A triple $(\Gamma_1, \Gamma_2, \tau)$ satisfying the above conditions is called *admissible*.

Let $(\Gamma_1, \Gamma_2, \tau)$ be an admissible triple. The continuous parameter on which the solutions depend is a tensor $r_0 \in \mathfrak{h} \otimes \mathfrak{h}$ satisfying the system of equations

(4.2.3) $$r_0^{12} + r_0^{21} = t_0,$$
(4.2.4) $$(\tau(\alpha) \otimes 1 + 1 \otimes \alpha)(r_0) = 0, \quad \alpha \in \Gamma_1.$$

Choose a system of Chevalley–Weyl generators $X_\alpha, X_{-\alpha}, H_\alpha$ (where α is a positive root) so that $(X_\alpha, X_{-\alpha}) = 1$. Denote by $\mathfrak{s}_i$ the Lie subalgebra of $\mathfrak{g}$ generated by $X_\alpha, X_{-\alpha}, H_\alpha$ where $\alpha \in \Gamma_i$ ($i = 1, 2$). One can easily see that

$$\mathfrak{s}_i = \sum_{\alpha \in \Gamma_i} \mathbb{C} H_\alpha \oplus \sum_{\alpha \in \hat{\Gamma}_i} \mathfrak{g}^\alpha$$

where $\hat{\Gamma}_i$ is the set of roots such that only the elements of Γ_i can appear in their expansions into a sum of simple roots, and $\mathfrak{g}^\alpha$ is the root subspace corresponding to α (that is, spanned by X_α).

It is easy to see that the relations between $X_\alpha, X_{-\alpha}, H_\alpha$ ($\alpha \in \Gamma_1$) are the same as those between $X_{\tau(\alpha)}, X_{-\tau(\alpha)}, H_{\tau(\alpha)}$. Therefore, there exists a unique isomorphism $\tilde{\tau} : \mathfrak{s}_1 \xrightarrow{\sim} \mathfrak{s}_2$ such that

$$\tilde{\tau}(X_{\pm\alpha}) = X_{\pm\tau(\alpha)}, \quad \tilde{\tau}(H_\alpha) = H_{\tau(\alpha)}.$$

Finally, we introduce a partial order $\prec$ on the set of roots so that $\alpha \prec \beta$ if there is $k \in \mathbb{N}$ such that $\beta = \tau^k(\alpha)$. Note that if $\alpha \prec \beta$, then $\alpha \in \hat{\Gamma}_1$ and $\beta \in \hat{\Gamma}_2$.

THEOREM 4.2.2 (Belavin, Drinfeld). (1) *In the above setting, let $r_0 \in \mathfrak{h} \otimes \mathfrak{h}$ satisfy the system (4.2.3), (4.2.4). Then the tensor*

$$r = r_0 + \sum_{\alpha > 0} X_\alpha \otimes X_{-\alpha} + \sum_{\substack{\alpha, \beta > 0 \\ \alpha \prec \beta}} (X_\alpha \otimes X_{-\beta} - X_{-\beta} \otimes X_\alpha) \tag{4.2.5}$$

is a solution to the system (4.2.1), (4.2.2).

(2) *Any solution to the system (4.2.1), (4.2.2) can be transformed into a solution of the form (4.2.5) by a suitable Lie algebra automorphism and multiplication by a constant.*

We give a proof in a series of exercises and comments to them. But before doing that, let us illustrate the statements of the theorem with some examples.

EXAMPLE 4.2.3. Suppose that $\mathfrak{g} = \mathfrak{sl}(2, \mathbb{C})$. In this case, the set of simple roots Γ consists of just one element. Hence, the only admissible triple is the one with both Γ_1 and Γ_2 being empty.

On the other hand, the only $r_0 \in \mathfrak{h} \otimes \mathfrak{h}$ such that

$$r_0^{12} + r_0^{21} = t_0 = \frac{1}{2} H \otimes H$$

is given by

$$r_0 = \frac{1}{4} H \otimes H.$$

Therefore, Theorem 4.2.2 gives us the solution (4.1.6).

However, this is not the only quasi-triangular Lie bialgebra structure on $\mathfrak{sl}(2, \mathbb{C})$. There exist some triangular ones described in Example 4.3.6.

EXAMPLE 4.2.4. As is easy to see, the standard Lie bialgebra structure on $\mathfrak{g}$ is a special case of the Belavin–Drinfeld list, with r given by (4.1.7).

PROOF. The rest of the section is devoted to the proof of the Belavin–Drinfeld Theorem 4.2.2. It is well know that the linear space of $\mathfrak{g}$-invariant elements in $\bigwedge^3 \mathfrak{g}$ is one-dimensional if $\mathfrak{g}$ is simple. Let $\Omega \in \left(\bigwedge^3 \mathfrak{g}\right)^{inv}$ be the only element of norm 1 (as the scalar product can be extended to $\mathfrak{g}^{\otimes 3} \supset \bigwedge^3 \mathfrak{g}$). Then the fact that the Lie bialgebra $(\mathfrak{g}, r)$ is quasi-triangular can be rewritten in the form

$$\langle r, r \rangle = \lambda \Omega, \tag{4.2.6}$$

where $\lambda \in \mathbb{C}$. Consider the case $\lambda \neq 0$.

EXERCISE 4.2.5. Consider the linear map $f : \mathfrak{g} \to \mathfrak{g}$ corresponding to $\tilde{r} \in \mathfrak{g} \wedge \mathfrak{g} \subset \mathfrak{g} \otimes \mathfrak{g}$ with respect to the isomorphism $\mathfrak{g}^* \simeq \mathfrak{g}$ given by the scalar product. Then the following statements are true:

(1) f is skew-symmetric with respect to the scalar product.

(2) The quasi-triangularity equation (4.2.6) is equivalent to the following identity, valid for any $x, y \in \mathfrak{g}$:

$$[f(x), f(y)] - f([f(x), y]) - f([x, f(y)]) = \lambda [x, y]. \tag{4.2.7}$$

For any complex number μ, consider

$$\mathfrak{g}_\mu = \{x \in \mathfrak{g} \mid (f - \mu)^k x = 0 \text{ for some } k > 0\}.$$

EXERCISE 4.2.6. (1) $[\mathfrak{g}_\mu, \mathfrak{g}_\nu] \subset \mathfrak{g}_\sigma$ for any $\mu \neq -\nu$, where $\sigma = \frac{\mu\nu - \lambda}{\mu + \nu}$.
(2) $\mathfrak{g}_\mu$ is orthogonal to $\mathfrak{g}_\nu$ for $\mu \neq \nu$.
(3) $[\mathfrak{g}_\mu, \mathfrak{g}_{-\mu}] = 0$ if $\mu^2 \neq -\lambda$.
(4) $\mathfrak{g}_a$ and $\mathfrak{g}_{-a}$ are Lie subalgebras in $\mathfrak{g}$ isotropic with respect to the scalar product, if $a^2 = -\lambda$.
(5) $\mathfrak{g}^{(a)} \stackrel{\text{def}}{=} \bigoplus_{\mu \neq \pm a} \mathfrak{g}_\mu$ is a Lie subalgebra in $\mathfrak{g}$, if $a^2 = -\lambda$.
(6) $\mathfrak{g}_{\pm a} \oplus \mathfrak{g}^{(a)}$ is a Lie subalgebra in $\mathfrak{g}$ containing $\mathfrak{g}_{\pm a}$ as an ideal, if $a^2 = -\lambda$.

Let $f^\pm = f \pm a$, where $a^2 \neq -\lambda$.

EXERCISE 4.2.7. (1) $f^\pm$ is invertible in $\mathfrak{g}_{\pm a} \oplus \mathfrak{g}^{(a)}$.
(2) It follows from (4.2.7) that

$$(4.2.8) \qquad f^+\left([f^-(x), f^-(y)]\right) = f^-\left([f^+(x), f^+(y)]\right)$$

for any $x, y \in \mathfrak{g}$.

COROLLARY 4.2.8. (1) *The linear map $\psi = f^+ \circ (f^-)^{-1}$ is a well-defined automorphism of the Lie algebra $\mathfrak{g}^{(a)}$.*
(2) *The number 1 is not an eigenvalue of ψ.*

PROOF. The first statement follows immediately from Exercise 4.2.7. To prove (2), assume that $\psi(x) = x$ for some $x \in \mathfrak{g}^{(a)}$. Then $f^+(x) = f^-(x)$, hence $2ax = 0$. Since $a^2 = -\lambda \neq 0$, we see that $x = 0$. □

EXERCISE 4.2.9. (1) Let $\mathfrak{b}$ be a finite-dimensional complex Lie algebra, and $\psi: \mathfrak{b} \to \mathfrak{b}$ a Lie algebra automorphism. Then ψ has 1 as an eigenvalue.
(2) If $\psi: \mathfrak{b} \to \mathfrak{b}$ is an automorphism of a finite-dimensional complex Lie algebra, and 1 is not an eigenvalue of ψ, then $\mathfrak{b}$ is solvable.

COROLLARY 4.2.10. *The Lie algebras $\mathfrak{g}_a$, $\mathfrak{g}_{-a}$, and $\mathfrak{g}_{\pm a} \oplus \mathfrak{g}^{(a)}$ are all solvable for $a^2 = -\lambda$. Moreover, $\mathfrak{g} = \mathfrak{g}_a \oplus \mathfrak{g}^{(a)} \oplus \mathfrak{g}_{-a}$.*

PROOF. The second statement follows from the definitions. The first one folows from Exercise 4.2.9 and Corollary 4.2.8. □

Since the Lie algebras $\mathfrak{g}_a \oplus \mathfrak{g}^{(a)}$ and $\mathfrak{g}_{-a} \oplus \mathfrak{g}^{(a)}$ are solvable, they are contained in some Borel subalgebras $\mathfrak{b}_+$ and $\mathfrak{b}_-$ respectively. By Corollary 4.2.10, we have $\mathfrak{b}_+ + \mathfrak{b}_- = \mathfrak{g}$. Define a Cartan subalgebra in $\mathfrak{g}$ by $\mathfrak{h} = \mathfrak{b}_+ \cap \mathfrak{b}_-$. Let $\mathfrak{n}_+$ (resp. $\mathfrak{n}_-$) be a nilpotent radical in $\mathfrak{b}_+$ (resp. $\mathfrak{b}_-$).

EXERCISE 4.2.11. The decomposition $\mathfrak{b}_\pm = \mathfrak{h} \oplus \mathfrak{n}_\pm$ holds, and $\mathfrak{n}_\pm$ is orthogonal to $\mathfrak{b}_\pm$.

COROLLARY 4.2.12. *The Lie algebra $\mathfrak{b}_+$ (resp. $\mathfrak{b}_-$) is the only Borel subalgebra in $\mathfrak{g}$ which contains $\mathfrak{g}_a \oplus \mathfrak{g}^{(a)}$ (resp. $\mathfrak{g}_{-a} \oplus \mathfrak{g}^{(a)}$).*

PROOF. We know that $\mathfrak{b}_+$ is the only Borel subalgebra containing the nilpotent Lis subalgebra $\mathfrak{n}_+$. We also have that $\mathfrak{g}_a \oplus \mathfrak{g}^{(a)}$ is orthogonal to $\mathfrak{g}_a$. Since $\mathfrak{g}_a \oplus \mathfrak{g}^{(a)} \subset \mathfrak{b}_+$, we see that $\mathfrak{n}_+ \subset \mathfrak{g}_a$. Similarly, one can show that $\mathfrak{n}_- \subset \mathfrak{g}_{-a}$. This proves the corollary. □

Let Δ be the set of roots of $\mathfrak{g}$ relative to $\mathfrak{h}$. We denote by $\mathfrak{g}^\alpha$ ($\alpha \in \Delta$) the corresponding root spaces. The choice of $\mathfrak{b}_+$ and $\mathfrak{b}_-$ defines the set of positive

roots Δ_+ and the set of negative roots Δ_-. We have
$$\mathfrak{b}_\pm = \mathfrak{h} \oplus \left(\bigoplus_{\alpha \in \Delta_+} \mathfrak{g}^{\pm\alpha} \right).$$

EXERCISE 4.2.13. Prove that $\mathfrak{b}_\pm$, $\mathfrak{n}_\pm$, and $\mathfrak{h}$ all are invariant with respect to f.

Let $\mathfrak{g}_\pm$ denote the image of $f^\pm$.

EXERCISE 4.2.14. (1) $\mathfrak{g}_\pm$ is a Lie subalgebra in $\mathfrak{g}$.
(2) $\mathfrak{g}_{\pm a} \oplus \mathfrak{g}^{(a)} \subset \mathfrak{g}_\pm$ (in particular, $\mathfrak{n}_\pm \subset \mathfrak{g}_\pm$).
(3) $\mathfrak{h} \oplus \mathfrak{g}_\pm$ is a parabolic subalgebra in $\mathfrak{g}$ which contains $\mathfrak{b}_\pm$.

Let $\Pi \subset \Delta_+$ be the set of simple roots.

EXERCISE 4.2.15. There exist subsets $\Gamma_\pm \subset \Pi$ such that $\mathfrak{g}^\pm$ defined as the images of $f^\pm \pm a$ admit the decompositions
$$\mathfrak{n}_{\Gamma_\pm} \oplus \sum_{\alpha \in \hat{\Gamma}_\pm} \left(\mathfrak{g}^\alpha \oplus \mathfrak{g}^{-\alpha} \oplus [\mathfrak{g}^\alpha, \mathfrak{g}^{-\alpha}] \right) \oplus V^\pm,$$
where:
- $\hat{\Gamma}_\pm$ is the set of positive roots which can be represented as integer linear combinations of the simple roots from $\Gamma_\pm$,
- $\mathfrak{n}_{\Gamma_\pm} = \bigoplus_{\alpha \in \Delta_+ \setminus \hat{\Gamma}_\pm} \mathfrak{g}^{\pm\alpha}$,
- $V^\pm \subset \mathfrak{h}$ is the subspace in $\left(\bigoplus_{\alpha \in \hat{\Gamma}_\pm} H_\alpha \right)^\perp$, where H_α consists of the coroots corresponding to α, and we take the orthogonal complement with respect to the scalar product. (Note that $(V^\pm)^\perp \subset V^\pm$.)

COROLLARY 4.2.16. *We have* $\operatorname{Ker}(f \pm a) = \mathfrak{n}_{\Gamma_\mp} \oplus (V^\mp)^\perp$.

PROOF. Note that $\operatorname{Ker}(f \pm a) = \operatorname{Im}(f \mp a)^\perp$ and use Exercise 4.2.15. □

EXERCISE 4.2.17. The equation (4.2.8) implies that the map $(f - a)(f + a)^{-1}$ induces a Lie algebra homomorphism $\theta : \mathfrak{g}^+/(\mathfrak{g}^+)^\perp \to \mathfrak{g}^-/(\mathfrak{g}^-)^\perp$.

By Exercise 4.2.15, we have
$$\mathfrak{g}^\pm/(\mathfrak{g}^\pm)^\perp = \bigoplus_{\alpha \in \hat{\Gamma}_\pm} \mathfrak{g}^\alpha \oplus \mathfrak{g}^{-\alpha} \oplus [\mathfrak{g}^\alpha, \mathfrak{g}^{-\alpha}] \oplus V^\pm/(V^\pm)^\perp.$$

Thus, we see that there is a map $\tau : \Gamma_+ \to \Gamma_-$ such that $(\tau(\alpha), \tau(\beta)) = (\alpha, \beta)$ for any $\alpha, \beta \in \Gamma_+$. Choose a Cartan basis $\{H_\alpha\}$, $\{E_\alpha\}$ ($\alpha \in \Delta_+$) in $\mathfrak{g}$ such that $H_\alpha \in \mathfrak{h}$, $E_\alpha \in \mathfrak{g}^\alpha$, and $\theta(H_\alpha) = H_{\tau(\alpha)}$, $\theta(E_\alpha) = E_{\tau(\alpha)}$.

EXERCISE 4.2.18. The linear map $1 - \psi$ on $\mathfrak{g}^\pm$ is invertible if and only if for any $\alpha \in \Gamma_+$, there is a positive integer k such that $\alpha, \tau(\alpha), \ldots, \tau^{k-1}(\alpha) \in \Gamma_+$ and $\tau^k(\alpha) \notin \Gamma_+$.

Given τ, we can write $f = a(1 - \psi)^{-1}(1 + \psi) = a(1 + \psi + \psi^2 + \cdots)(1 + \psi)$. This implies
$$f(E_\gamma) = aE_\gamma, \quad \gamma \in \Delta_+ \setminus \hat{\Gamma}_+,$$
$$f(E_\alpha) = a\left(E_\alpha + 2 \sum_{\beta \succ \alpha} E_\beta \right), \quad \alpha \in \hat{\Gamma}_+.$$

Here the notation $\beta \succ \alpha$ for $\alpha, \beta \in \hat{\Gamma}_+$ means that there exists $k > 1$ such that $\tau^k(\alpha) = \beta$.

EXERCISE 4.2.19. The linear map f is completely determined by its restriction to $\mathfrak{n}_-$.

LEMMA 4.2.20. We have that $(f - a)(H_\alpha) = (f + a)(H_{\tau(\alpha)})$ for any $\alpha \in \Gamma_+$.

PROOF. Since $\theta(H_\alpha) = H_{\tau(\alpha)}$, we see that $(f - a)(x) = H_{\tau(\alpha)} + y$, where x is such that $(f + a)(x) = H_\alpha$ and $y \in \text{Ker}(f + a)$. Note also that the restriction of f to $\mathfrak{h}$ is skew-symmetric. □

EXERCISE 4.2.21. (1) Prove that a solution $r \in \mathfrak{g} \wedge \mathfrak{g}$ to (4.2.6) is given by (4.2.5) with $\Gamma_1 = \Gamma_+$ and $\Gamma_2 = \Gamma_-$.
(2) Conversely, prove that if r is given by (4.2.5), then the map f it defines satisfies (4.2.7), and, therefore, r is a solution to (4.2.6).

This completes the proof of the Belavin–Drinfeld theorem. □

4.3. Frobenius Lie algebras and CYBE. The Belavin–Drinfeld theorem does not include the case of triangular Lie bialgebras. We show that the classification problem for triangular Lie bialgebra structures contains a subproblem of describing all Frobenius Lie subalgebras, and give some examples.

Recall that given a finite-dimensional vector space V with nondegenerate bilinear form $\langle , \rangle$, a tensor $T \in V \otimes V$ defines a bilinear form B_T on V by

$$B_T(v_1, v_2) = \langle T, v_1 \otimes v_2 \rangle.$$

Then the tensor T is called *nondegenerate* if the corresponding bilinear form is nondegenerate.

REMARK 4.3.1. Let $\mathfrak{g}$ be a Lie algebra with nondegenerate invariant bilinear form $\langle , \rangle$, and $r \in \mathfrak{g} \wedge \mathfrak{g}$. We can always choose a Lie subalgebra $\mathfrak{p} \subset \mathfrak{g}$ such that $r \in \mathfrak{p} \wedge \mathfrak{p}$ and r is nondegenerate in $\mathfrak{p} \wedge \mathfrak{p}$.

It is clear that $r \in \mathfrak{g} \wedge \mathfrak{g}$ defines a triangular Lie bialgebra structure on $\mathfrak{g}$ if and only if $r \in \mathfrak{p} \wedge \mathfrak{p}$ defines a triangular Lie bialgebra structure on $\mathfrak{p}$. In other words, one can reduce the problem of finding solutions to CYBE (4.1.3) to the case where r is nondegenerate.

PROPOSITION 4.3.2. *Let $\mathfrak{p}$ be a Lie algebra, $r \in \mathfrak{p} \wedge \mathfrak{p}$ a nondegenerate tensor, and $B = B_r$ the corresponding bilinear form on $\mathfrak{p}$. Then r is a classical r-matrix for a triangular Lie bialgebra structure on $\mathfrak{p}$ if and only if B is a 2-cocycle on $\mathfrak{p}$, i.e. if for any $x, y, z \in \mathfrak{p}$ we have*

(4.3.1) $$B([x, y], z) + B([y, z], x) + B([z, x], y) = 0.$$

PROOF. Suppose that r is a classical r-matrix for a triangular Lie bialgebra structure on $\mathfrak{p}$. Then the cobracket $\varphi : \mathfrak{p} \to \mathfrak{p} \wedge \mathfrak{p}$ is given by the formula

(4.3.2) $$\varphi(a) = [a \otimes 1 + 1 \otimes a, r].$$

Then we have that

$$\begin{aligned} B([x, y], z) &= \langle r, [x, y] \otimes z \rangle = \langle (\varphi \otimes \text{id})(r), x \otimes y \otimes z \rangle \\ &= \langle [r^{13} + r^{23}, r^{12}], x \otimes y \otimes z \rangle. \end{aligned}$$

By performing the cyclical permutations and adding the summands together, we get the CYBE $\langle r, r \rangle = 0$. Therefore, (4.3.1) is satisfied.

On the other hand, suppose that the bilinear form $B = B_r$ satisfies (4.3.1). Then if we define the cobracket by (4.3.2), we get that r satisfies the CYBE and, therefore, defines a triangular Lie bialgebra structure on $\mathfrak{p}$. □

By the above proposition, any pair $(\mathfrak{p}, B)$, where $\mathfrak{p}$ is a Lie subalgebra of $\mathfrak{g}$ and B a nondegenerate 2-cocycle on $\mathfrak{p}$, gives us a solution to CYBE (4.1.3), i.e. a triangular Lie bialgebra structure on $\mathfrak{g}$.

In what follows we restrict ourselves to the special case when B is a coboundary. This means that there exists a linear functional $l \in \mathfrak{p}^*$ such that

$$B(x, y) = l([x, y])$$

for any $x, y \in \mathfrak{p}$.

DEFINITION 4.3.3. A Lie algebra $\mathfrak{p}$ is called *Frobenius Lie algebra* if there exists $l \in \mathfrak{p}^*$ such that the bilinear form

$$\langle a, b \rangle = l([a, b])$$

is nondegenerate on $\mathfrak{p}$.

REMARK 4.3.4. By Proposition 4.3.2, any Frobenius Lie subalgebra of $\mathfrak{g}$ yields a triangular Lie bialgebra structure on $\mathfrak{g}$.

EXERCISE 4.3.5. Show that given a Frobenius Lie algebra $\mathfrak{p}$, a linear functional $l \in \mathfrak{p}^*$ satisfting the condition of Definition 4.3.3 is unique up to conjugation.

EXAMPLE 4.3.6. Suppose that $\mathfrak{g} = \mathfrak{sl}(2, \mathbb{C})$. Let $\mathfrak{b}_\pm \subset \mathfrak{g}$ be the Borel subalgebra generated by H and $X^\pm$. It is easy to check that the tensor

$$(4.3.3) \qquad r = H \wedge X^\pm$$

satisfies CYBE (4.1.3). The corresponding bilinear form B on $\mathfrak{b}_\pm$ is given by

$$B\left(aH + bX^\pm, cH + dX^\pm\right) = \det \begin{pmatrix} a & b \\ c & d \end{pmatrix}.$$

Let $l \in \mathfrak{b}_\pm^*$ be a linear functional given by

$$l\left(aH + bX^\pm\right) = \frac{b}{2}.$$

Then it is easy to see that $B(x, y) = l([x, y])$ for any $x, y \in \mathfrak{b}_\pm$. That is, $\mathfrak{b}_\pm$ is a Frobenius Lie algebra. The corresponding triangular Lie bialgebra structure on $\mathfrak{sl}(2, \mathbb{C})$ is given by the classical r-matrix (4.3.3).

EXERCISE 4.3.7. Show that the only nontrivial Lie bialgebra structures on $\mathfrak{sl}(2, \mathbb{C})$ up to isomorphism are those given in Example 4.2.3 and Example 4.3.6. Hint: use the following proposition.

PROPOSITION 4.3.8. *If $\mathfrak{g}$ is a semisimple Lie algebra over $\mathbf{k}$, then $\mathfrak{g}$ is not Frobenius.*

PROOF. Assume that $\mathfrak{g}$ is a Frobenius Lie algebra. Then for any $x, y, z \in \mathfrak{g}$, we have

$$l([[z, x], y]) = -l([x, [z, y]])$$

for some nontrivial $l \in \mathfrak{g}^*$. By the Jacobi identity, we get that

$$l([[x, y], z]) = 0.$$

Since $[\mathfrak{g},\mathfrak{g}] = \mathfrak{g}$ for any semisimple Lie algebra $\mathfrak{g}$, we get that $\mathrm{ad}^*_z(l) = 0$ for any $z \in \mathfrak{g}$. But there are no nontrivial $\mathfrak{g}$-invariant functionals on a semisimple Lie algebra. This contradiction shows that $\mathfrak{g}$ is not Frobenius. $\square$

We observe that the classification of triangular Lie bialgebra structures on $\mathfrak{g}$ contains as a subproblem the description of all Frobenius Lie subalgebras. Even though $\mathfrak{g}$ itself may not be Frobenius (for instance, if it is compact or complex simple), it still may contain Frobenius subalgebras. One example was given in Example 4.3.6. Here is a generalization.

EXERCISE 4.3.9. Let $\mathfrak{p}$ be the Lie algebra of complex $n \times n$ matrices A such that all the elements of the last k rows are equal to zero. Then $\mathfrak{p}$ is Frobenius if and only if k divides n. In this case the corresponding linear functional $l \in \mathfrak{p}^*$ can be given by

$$l(A) = \sum_{i=1}^{n-1} a_{i,i+k},$$

for any $A = (a_{ij})_{i,j=1}^n$. The corresponding classical r-matrix is as follows:

$$r = \sum_{i,j=1}^{k} \sum_{(a,b,c,d) \in \Omega} e_{i+ka,j+kb} \wedge e_{j+kc,i+kd}$$

where e_{ab} is the matrix whose only nonzero entry is in the ath row and bth column and is equal to 1,

$$\Omega = \{(a,b,c,d) \in \mathbb{Z}^4 \mid b+d-a-c = 1, 0 \leq b \leq a < m-1,$$
$$b \leq c \leq m-1, 0 \leq d \leq m\}, \qquad m = \frac{n}{k} \in \mathbb{Z}.$$

The embedding $\mathfrak{p} \hookrightarrow \mathfrak{sl}(n,\mathbb{C})$ such that

$$A \mapsto A - \mathrm{tr}\, A$$

provides a solution $(f \otimes f)(r) \in \mathfrak{sl}(n,\mathbb{C}) \wedge \mathfrak{sl}(n,\mathbb{C})$ to CYBE (4.1.3), which defines a triangular Lie bialgebra structure on $\mathfrak{sl}(n,\mathbb{C})$.

5. Compact Poisson Lie groups

5.1. Quasi-triangular compact Poisson Lie groups. In the present subsection we give a complete classification of Poisson Lie group structures on compact Lie groups.

Let G be a connected and simply connected complex simple Lie group, and $\mathfrak{g}$ the Lie algebra of G. It is well known that there exists a unique (up to inner automorphism) compact real form of G. Denote it by K, and its Lie algebra by $\mathfrak{k}$.

This Lie algebra can be described as follows. Let ω_0 be an antilinear involutive Lie algebra automorphism given by

(5.1.1) $\qquad \omega_0\left(X_i^{\pm}\right) = -X_i^{\mp}, \quad \omega_0(H_i) = -H_i, \quad (i=1,2,\ldots,n).$

Then the Lie algebra $\mathfrak{k} \subset \mathfrak{g}$ coincides with the set of fixed points of ω_0:

$$\mathfrak{k} = \{a \in \mathfrak{g} \mid \omega_0(a) = a\}.$$

The following theorem gives a classification of Poisson Lie group structures on K. As usual, we suppose that a decomposition (2.1.3) of $\mathfrak{g}$ is fixed.

THEOREM 5.1.1. (1) *If $\varphi : \mathfrak{k} \to \mathfrak{k} \wedge \mathfrak{k}$ defines a Lie bialgebra structure on $\mathfrak{k}$, then $\varphi = \partial r$ for some $r \in \mathfrak{k} \wedge \mathfrak{k}$.*
(2) *Let*

(5.1.2) $$r_0 = \frac{i}{2} \sum_{\alpha > 0} X_{-\alpha} \wedge X_{\alpha}.$$

Then for any $r' \in \mathfrak{k} \wedge \mathfrak{k}$ that satisfies MCYBE (4.1.2) there exists a Lie algebra automorphism of $\mathfrak{k}$ which transforms r' into

(5.1.3) $$r(a, u) = ar_0 + u$$

for some $a \in \mathbb{R}$ and $u \in \mathfrak{t} \wedge \mathfrak{t}$, where $\mathfrak{t} = \mathfrak{k} \cap \mathfrak{h}$.

In particular, any Poisson Lie group structure on K is isomorphic up to inner automorphism to the one given by the classical r-matrix $r(a, u)$. We denote the corresponding Poisson Lie group by $K(a, u)$.

PROOF. The statement (1) follows from the fact that $H^1(\mathfrak{k}, \mathfrak{k} \wedge \mathfrak{k}) = 0$ for any simple Lie algebra $\mathfrak{k}$. To prove (2), we note that by Proposition 4.3.8, there is no Frobenius Lie bialgebra structure on $\mathfrak{k}$ (since $\mathfrak{k}$ is a real simple Lie algebra). It follows from the Belavin–Drinfeld Theorem 4.2.2 that any other quasi-triangular Lie bialgebra structure on $\mathfrak{k}$—that is, an ω_0-invariant quasi-triangular Lie bialgebra structure on G—is given, up to a Lie algebra automorphism, by a classical r-matrix of the form (5.1.3). The same is true for any Lie subalgebra of $\mathfrak{k}$ (which is automatically compact, too). This proves the proposition. □

REMARK 5.1.2. We will be particularly interested in the two families $K(1, u)$ and $K(0, u)$ of Poisson Lie group structures on K. We leave it as an exercise to generalize the results of the rest of the section to the case $a \neq 0, 1$. In fact, the picture of symplectic leaves is the same for any $a \neq 0$.

EXAMPLE 5.1.3. If $G = SL_2(\mathbb{C})$, then $K = SU(2)$. In this case $\mathfrak{t} \wedge \mathfrak{t} = 0$, so that there is only one nontrivial quasi-triangular Poisson Lie group structure on $SU(2)$ up to inner automorphism.

5.2. Symplectic leaves and Bruhat decomposition. Now we want to describe the symplectic leaves in the Poisson Lie group $K(a, u)$. It turns out that they are related to the Bruhat decomposition.

Let $\mathfrak{b}_{\pm} = \mathfrak{n}_{\pm} \oplus \mathfrak{h}$ be the Borel subalgebras of $\mathfrak{g}$, and $B_{\pm}$ the corresponding Borel subgroups of G. Recall that the Weyl group W is generated by the simple reflections $s_i : \mathfrak{h}^* \to \mathfrak{h}^*$ given by

$$s_i(\lambda) = \lambda - \frac{2(\lambda, \alpha_i)}{(\alpha_i, \alpha_i)} \alpha_i$$

where $\alpha_1, \ldots, \alpha_r$ are the simple roots, and $r = \dim \mathfrak{h}$ is the rank of $\mathfrak{g}$.

It is known that one can choose a finite covering $\widetilde{W} \to W$ such that $\widetilde{W}$ is a subgroup in K. Suppose that we have chosen a representative in $\widetilde{W}$ for each $w \in W$ which we denote by the symbol $\dot{w}$. The following result is well known (cf. [**Ste72**]).

PROPOSITION 5.2.1 (Bruhat decomposition). (1) *The following decomposition of G holds:*

(5.2.1) $$G = \bigcup_{w \in W} B_{\pm} \dot{w} B_{\pm}.$$

It is is called the Bruhat decomposition *of G.*

(2) *The following decomposition of K holds:*

(5.2.2) $$K = \bigcup_{w \in W} K_w \quad \text{where} \quad K_w = B_\pm \dot{w} B_\pm \cap K.$$

It is is called the Bruhat decomposition *of K.*

Let $\mathfrak{t} = \mathfrak{k} \cap \mathfrak{h}$. Consider the maximal torus $T = \exp \mathfrak{t}$ in K and the corresponding flag manifold $X = K/T \simeq G/B_\pm$. By Proposition 5.2.1, we have

(5.2.3) $$X = \bigcup_{w \in W} X_w \quad \text{where} \quad X_w \simeq B_\pm \dot{w} B_\pm / B_\pm$$

is the so-called *Schubert cell* of X corresponding to $w \in W$. It is well known that X_w is naturally isomorphic to $\mathbb{C}^{l(w)}$, where $l(w)$ is the length of w (cf. [**Ste72**]).

THEOREM 5.2.2. (1) *Each symplectic leaf $S \subset K(1,0)$ lies in K_w for some $w \in W$, and each Bruhat class K_w is a Poisson submanifold in $K(1,0)$.*

(2) *Each symplectic leaf $S \subset K(1,0)$ is of the form*

$$S = \Sigma_w \cdot t,$$

where $t \in T$ and Σ_w is the symplectic leaf passing through $\dot{w} \in K$.

PROOF. By Example 3.3.8, we have that $G \simeq \mathcal{D}(K)$, so that $K^* \simeq AN_+$. By Theorem 3.4.14, the symplectic leaf in K passing through $g \in K$ is naturally diffeomorphic to $AN_+ g AN_+ / AN_+$. It is easy to see that it lies in $K_w = B_+ \dot{w} B_+ / B_+$ for some $w \in W$. In particular, the symplectic leaf Σ_w passing through $\dot{w}$ lies in K_w.

On the other hand, $g \mapsto gt$ is a Poisson map for any $t \in T$. Therefore, $\Sigma_w t$ is a symplectic leaf for any $t \in T$. Since $\Sigma_w = AN_+ \dot{w} AN_+ / AN_+$, the Bruhat class K_w is the union of all symplectic leaves of the form $\Sigma_w t$ ($t \in T$) and, therefore, a Poisson submanifold in K. In particuar, T is the union of zero-dimensional symplectic leaves. □

The Bruhat decomposition is related to the decomposition of $K(a, 0)$ into symplectic leaves. Since everything is the same for any $a \neq 1$, we can assume that $a = 1$ without loss of generality. In what follows we will use the simpler notation K in place of $K(1,0)$. This particular Poisson Lie group structure on K is called *standard*.

COROLLARY 5.2.3. (1) *There is a unique Poisson manifold structure on the flag manifold X such that the projection $K \to X$ is a Poisson map.*

(2) *The Schubert cell X_w is a symplectic leaf in X for any $w \in W$. In other words, the decomposition (5.2.3) is the decomposition of X into the union of symplectic leaves.*

EXAMPLE 5.2.4. Suppose that $K = SU(2)$. In this case W is the group of two elements, the unit element e and w_0; the latter can be represented in $SU(2)$ by $w_0 = \begin{pmatrix} 0 & 1 \\ -1 & 0 \end{pmatrix}$. Therefore, there are two series of symplectic leaves in $SU(2)$ parameterized each by the points of $T \simeq S^1$. One of the series consists of the zero-dimensional leaves which are the points of T, while the other one consists of the two-dimensional symplectic leaves of the form

(5.2.4) $$S_t = \left\{ \begin{pmatrix} a & -\bar{b} \\ b & \bar{a} \end{pmatrix} \in SU(2) \mid \arg b = \arg t \right\},$$

where $t \in \mathbb{C}$, $|t| = 1$.

The flag manifold X in this example is isomorphic to $\mathbb{C}P^1$ and can be thought of as the Riemann sphere. The projection images of Σ_e and Σ_{w_0} in X are the Schubert cells $\{\infty\}$ and $\mathbb{C} \simeq \mathbb{C}P^1 \setminus \{\infty\}$.

We give another proof of Theorem 5.2.2 and Corollary 5.2.3. It is based on the following proposition. Let i be a vertex of the Dynkin diagram of $\mathfrak{g}$, and $\tau_i : SU(2) \hookrightarrow K$ the corresponding embedding of Lie groups. In fact, τ_i is an embedding of Poisson Lie groups. Let $s_i \in W$ be the simple reflection corresponding to i. As is well known, the simple reflections s_i generate W. Recall that an expression $w = s_{i_1} \cdots b s_{i_k}$ is called *reduced* if there is no other decomposition of w into a product of simple reflections which would be of length less than k.

PROPOSITION 5.2.5. (1) *The symplectic leaf Σ_{s_i} in K passing through s_i is two-dimensional and has the form*
$$\Sigma_{s_i} = \tau_i(S_{-1}),$$
where S_{-1} is given by (5.2.4).

(2) *Let $w \in W$ and $w = s_{i_1} \cdots b s_{i_k}$ be a reduced expression. Then the symplectic leaf Σ_w in K passing through w is given by*

(5.2.5) $$\Sigma_w = \Sigma_{s_{i_1}} \cdots b \Sigma_{s_{i_k}}.$$

PROOF. The statement (1) follows immediately from the fact that τ_i is an embedding of Poisson Lie groups. It is well known (cf. [**Ste72**]) that the symplectic leaf Σ_w, $w = s_{i_1} \cdots s_{i_k}$ defined in the proof of Theorem 5.2.2 coincides with the image in K of the submanifold
$$\Sigma_{s_{i_1}} \times \Sigma_{s_{i_1}} \times \cdots \times \Sigma_{s_{i_k}} \subset SU(2) \times SU(2) \times \cdots \times SU(2)$$
with respect to the Poisson morphism $(g_1, \ldots, g_k) \mapsto g_1 \cdots g_k$. This proves (2). □

REMARK 5.2.6. We see that an arbitrary symplectic leaf in K is either a product of two-dimensional leaves (all isomorphic to $S_{-1} \simeq \mathbb{C}$) or a point of the maximal torus T. This implies that $X_w \simeq \Sigma_w$ is isomorphic to an affine manifold. More precisely, by Lemma 5.2.5, $X_w \simeq \mathbb{C}^{l(w)}$.

We can similarly consider the case of the Poisson Lie group $K(a, u)$ when $a \neq 0$. Without loss of generality we can assume that $a = 1$ and denote $K(1, u)$ by K^u.

Let $\mathfrak{a} = \sqrt{-1}\mathfrak{t}$. Denote by $\check{u}$ the linear map from $\mathfrak{t}^*$ to $\mathfrak{t}$ induced by $u \in \mathfrak{t} \wedge \mathfrak{t}$ via the isomorphism $\mathfrak{t}^* \simeq \mathfrak{t}$ defined by the scalar product in $\mathfrak{g}$. Finally, let $\mathfrak{a}_u \stackrel{\text{def}}{=} (\sqrt{-1} + 2\check{u})\mathfrak{t}$.

PROPOSITION 5.2.7. *The Lie algebras $(\mathfrak{g}, \mathfrak{k}, \mathfrak{a}_u \oplus \mathfrak{n}_+)$ form a Manin triple, the invariant symmetric bilinear form on $\mathfrak{g}$ given by the imaginary part of the Killing form.*

PROOF. Since $\mathfrak{a}_{\mathbb{C}} = \mathfrak{a} \oplus \mathfrak{t}$ normalizes $\mathfrak{n}_+$, we have that $\mathfrak{a}_u \oplus \mathfrak{n}_+$ is a Lie subalgebra in $\mathfrak{g}$. We leave it as an exercise to the reader to show that it is isotropic with respect to the imaginary part of the Killing form. □

Now we can prove an analog of Theorem 5.2.2 for $u \neq 0$.

THEOREM 5.2.8. (1) *Each symplectic leaf S^u of K^u lies in K_w for some $w \in W$.*

(2) *If S_{tw}^u is a symplectic leaf in K^u passing through tw, where $t \in T$ and $w \in W$, then*
$$S_{tw}^u = T_w \cdot t \cdot \Sigma_w.$$
Here Σ_w is the same as in Theorem 5.2.2 and $T_w \subset T$ is given by
$$T_w = \exp\left((\tilde{u} - w^{-1}\tilde{u}w)\mathfrak{t}\right),$$
where $\tilde{u} : \mathfrak{t} \to \mathfrak{t}$ corresponds to $u \in \mathfrak{t} \wedge \mathfrak{t}$.

PROOF. We can identify the dual Poisson Lie group K_u^* with the subgroup $\exp(\mathfrak{a}_u \oplus \mathfrak{n}_+)$ in G. The generic element of K_u^* is of the form
$$g = n \exp\left((1 + 2i\tilde{u})a\right),$$
where $n \in N_+$ and $a \in \mathfrak{a}$. By Theorem 3.4.14, the symplectic leaf Σ_g^u in K_u passing through $g \in K_u$ is of the form $K_u^* g K_u^* / K_u^*$. In other words, we have
$$\Sigma_g^u = \{h \in K \mid \exists a, b \in K_u^* : ag = bh\}.$$

Since $N_+\dot{w} = \Sigma_w N_+$, we get that $\Sigma_{tw}^u = T_w t \Sigma_w$. In particular, any symplectic leaf lies in some Bruhat class K_w. □

REMARK 5.2.9. We see that when $u \neq 0$, the picture of symplectic leaves is more complicated. The leaves do not have to be isomorphic to affine manifolds anymore. For instance, it is a typical situation when there exists a symplectic leaf which is dense in K.

Finally, we consider the case $a = 0, u \neq 0$. Unlike the cases considered before, now the symplectic leaves are not related to the Bruhat decomposition.

Consider the cotangent bundle $T^*K \simeq K \times \mathfrak{k}^*$ as the semidirect product of the Lie groups K and $\mathfrak{k}^*$ with respect to the coadjoint action of K on $\mathfrak{k}^*$. That is, the following multiplication law holds:

(5.2.6) $$(g_1, \alpha_1) \cdot (g_2, \alpha_2) = (g_1 g_2, \alpha_2 + \text{Ad}^*_{g_2^{-1}} \alpha_1),$$

for any $g_i \in K, \alpha_i \in \mathfrak{k}^*, i = 1, 2$. The Lie algebra of this semidirect product is the vector space $\mathfrak{k} \oplus \mathfrak{k}^*$ equipped with the Lie bracket

(5.2.7) $$[a + \alpha, b + \beta] = [a, b] + \text{ad}^*_a \beta - \text{ad}^*_b \alpha,$$

for any $a, b \in \mathfrak{k}, \alpha, \beta \in \mathfrak{k}^*$. Here Ad^* stands for the coadjoint action of K on $\mathfrak{k}^*$ and ad^* for the corresponding action of $\mathfrak{k}$ on $\mathfrak{k}^*$.

PROPOSITION 5.2.10. (1) *The double Poisson Lie group $\mathcal{D}(K(0, u))$ is isomorphic to $T^*K \simeq K \times \mathfrak{k}^*$ with the multiplication given by (5.2.6). The double Lie bialgebra $\mathcal{D}(\mathfrak{k}(0, u))$ is isomorphic to $\mathfrak{k} \oplus \mathfrak{k}^*$ with the Lie bracket given by (5.2.7).*

(2) *The dual Poisson Lie group $K(0, u)^*$ is isomorphic to the subgroup*
$$K(0, u)^* \simeq \exp\left((1 + \tilde{u})\mathfrak{t}^* \oplus (\mathfrak{t}^\perp)^*\right),$$
where $\mathfrak{t}^\perp$ is the orthogonal complement to $\mathfrak{t}$ in $\mathfrak{k}$.

PROOF. It suffices to note that $\left(\mathfrak{k} \oplus \mathfrak{k}^*, \mathfrak{k}, (1 + \tilde{u})\mathfrak{t}^* \oplus (\mathfrak{t}^\perp)^*\right)$ is a Manin triple, with the invariant bilinear form on $\mathfrak{k} \oplus \mathfrak{k}^*$ given by
$$\langle a + \alpha, b + \beta \rangle = \langle a, b \rangle + \langle \alpha, \beta \rangle,$$
for any $a, b \in \mathfrak{k}, \alpha, \beta \in \mathfrak{k}^*$. The proposition is proved. □

Now we can use Theorem 3.4.14 to obtain the following description of the symplectic leaves in $K(0, u)$.

PROPOSITION 5.2.11. *The symplectic leaf in $K(0, u)$ passing through a point g is of the form*
$$\Sigma_g^{0,u} = \{\exp(\tilde{u}\lambda)g\exp(-\tilde{u}\operatorname{pr}_1\operatorname{Ad}_{g^{-1}}^*(\lambda,\zeta)) \mid \lambda \in \mathfrak{t}^*, \zeta \in (\mathfrak{t}^\perp)^*\},$$
where pr_1 is the projection from $\mathfrak{t}^* \oplus \mathfrak{t}^*$ to the first component.

PROOF. The proof once again repeats the arguments of the proofs of Theorem 5.2.2 and Proposition 5.2.8. We leave it as an exercise to the reader to do it in this case. □

REMARK 5.2.12. The reader familiar with the orbit method knows that there is an intimate relation between coadjoint orbits in $\mathfrak{g}^*$ and $U\mathfrak{g}$-modules. Later in the book we will see that there is a correspondence between the symplectic leaves in $K = K(1,0)$ and the irreducible $*$-representations of the quantized algebra of functions on K.

However, the correspondence between the leaves and the representations fails for $K(0, u)$ when $u \neq 0$. Moreover, the space of symplectic leaves is nonseparable in most cases. This indicates that the dual Poisson Lie group $K(0, u)^*$ behaves in a certain sense like a wild Lie group.

We will return to this topic and discuss it in detail in Part II of this book.

5.3. Symplectic leaves in simple complex Poisson Lie groups. Let G be a finite-dimensional simple complex Lie group, and $\mathfrak{g}$ the Lie algebra of G. Suppose that $\mathfrak{g}$ is equipped with the Lie bialgebra structure called standard and described in Example 2.3.7. It induces a Poisson Lie group structure on G which is also called standard. Our goal is to describe the symplectic leaves in G.

By Example 2.3.7, there is a Manin triple $(\mathfrak{g} \oplus \mathfrak{g}, \mathfrak{g}, \mathfrak{s})$ (for the notation see Example 2.3.7). We see that the dual Poisson Lie group $\mathcal{D}(G)$ is isomorphic to $G \times G$ as Lie group. It is clear also that $G \cdot G^*$ is dense and open in $\mathcal{D}(G)$, but does not necessarily coincide with $\mathcal{D}(G)$. Moreover, the multiplication map $m : G \times G^* \to G \cdot G^*$ is a diffeomorphism.

We conclude that the image $\tilde{G}$ of the quotient map $p : G \to \mathcal{D}(G)/G^*$ is dense and open in $\mathcal{D}(G)/G^*$. The following result is a consequence of Theorem 3.4.14 and Exercise 3.4.11, so we omit the proof.

LEMMA 5.3.1. (1) *Any symplectic leaf in G is a connected component of the fiber $p^{-1}(S)$, where S is a symplectic leaf in $\tilde{G}$.*

(2) *Any symplectic leaf in $\tilde{G}$ is of the form $\tilde{G} \cap G^*gG^*/G^*$ for some $g \in G \subset \mathcal{D}(G)$.*

Now we can continue in the same way as we did in the case of a compact Poisson Lie group, so we omit the details. The following Bruhat decomposition holds:
$$\mathcal{D}(G) = \bigcup_{(w_1,w_2)\in W\times W} P \cdot (w_1, w_2) \cdot P,$$
where $P = HG^*$ and $H = \exp\mathfrak{h}$ is the distinguished Cartan subgroup of G.

Consider the following sets:

$$
\begin{aligned}
C_{(w_1,w_2)} &= (G^* \cdot (w_1,w_2) \cdot G^*)/G^*, \\
B_{(w_1,w_2)} &= C_{(w_1,w_2)} \cap \tilde{G}, \\
A_{(w_1,w_2)} &= p^{-1}\left(\bigcup_{h \in H} hB_{(w_1,w_2)}\right).
\end{aligned}
$$

The proofs of the following two propositions are similar to those of the results on symplectic leaves in a compact Poisson Lie group (presented earlier in this section). We leave them to the reader as exercises.

PROPOSITION 5.3.2. (1) *Each symplectic leaf in $\tilde{G}$ is of the form $hB_{(w_1,w_1)}$ for some $h \in H$ and $(w_1, w_2) \in W \times W$.*

(2) *Each symplectic leaf in G is of the form $hA_{(w_1,w_2)}$ for some $h \in H$ and $(w_1, w_2) \in W \times W$.*

PROPOSITION 5.3.3. *Denote $s(w_1, w_2) = \mathrm{codim}_\mathfrak{h} \mathrm{Ker}(w_1 w_2^{-1} - 1)$. Then*

$$C_{(w_1,w_2)} \simeq H^{s(w_1,w_2)} \times \mathbb{C}^{l(w_1)+l(w_2)},$$

where $l(w)$ is the length of the element w of the Weyl group W.

By Proposition 5.3.2, the symplectic leaves in G are parameterized by the maximal torus H and the Cartesian "square" $W \times W$ of the Weyl group W. Recall that by Theorem 5.2.2, the symplectic leaves in the compact real form K of G are parameterized by the maximal compact torus $T = H \cap K$ and the Weyl group W. In fact, the symplectic leaves in K are the intersections of K and the symplectic leaves in G of the form $hA_{(w,w)}$, where $w \in W$.

EXERCISE 5.3.4. Show that the following is the full list of the symplectic leaves in $SL_2(\mathbb{C})$:

$$
\begin{aligned}
T_t &= \left\{\begin{pmatrix} t & 0 \\ 0 & t^{-1} \end{pmatrix}\right\}, \quad t \neq 0, \\
T_t A_{(e,w_0)} &= \left\{\begin{pmatrix} t & b \\ 0 & t^{-1} \end{pmatrix} \mid b \neq 0\right\}, \quad t \neq 0, \\
T_t A_{(w_0,e)} &= \left\{\begin{pmatrix} t & 0 \\ c & t^{-1} \end{pmatrix} \mid c \neq 0\right\}, \quad t \neq 0, \\
T_t A_{(w_0,w_0)} &= \left\{\begin{pmatrix} a & b \\ c & d \end{pmatrix} \mid b, c \neq 0, \frac{b}{c} = t^2\right\}, \quad t \neq 0.
\end{aligned}
$$

6. Poisson G-manifolds

6.1. Poisson G-manifolds and moment maps.

DEFINITION 6.1.1. Let G be a Poisson Lie group, and X a left (resp. right) G-manifold. We call X *left* (resp. *right*) *Poisson G-manifold* if the group action $G \times X \to X$ (resp. $X \times G \to X$) is a Poisson map. Also, in this case the action of G on X is called *Poisson action*.

EXAMPLE 6.1.2. Suppose that G is a Lie group, and $\mathfrak{g}$ the Lie algebra of G. We can think of G as a Poisson Lie group with the trivial Poisson bracket. Consider also the Poisson Lie group structure on the vector space $\mathfrak{g}^*$ given in Example 3.2.7. We know from Example 3.3.6 that these two form a pair of dual Poisson Lie groups.

It is easy to see that the coadjoint action of G on $\mathfrak{g}^*$ makes the latter a left Poisson G-space.

In particular, we see that any coadjoint orbit is a Poisson G-manifold (provided that G is equipped with the trivial Poisson Lie group structure).

The above fact can be generalized to the case of an arbitrary Poisson Lie group G. In order to do it, we have to replace $\mathfrak{g}^*$ by the dual Poisson Lie group G^* and the coadjoint action by the dressing action.

PROPOSITION 6.1.3. *Let G be a Poisson Lie group and G^* the Poisson Lie group dual to G. Suppose that the local left (resp. right) dressing action of G on G^* can be extended to a global one. Then G^* is a left (resp. right) Poisson G-manifold with respect to the global dressing action. Moreover, any orbit of the left (resp. right) dressing action is a left (resp. right) Poisson G-manifold.*

PROOF. By Definition 3.4.4, the left dressing action of $G \times G^* \to G^*$ is the composition of the multiplication $G^* \times G \to \mathcal{D}(G)$ in $\mathcal{D}(G)$ and the natural projection $\mathcal{D}(G) \to \mathcal{D}(G)/G \simeq G^*$. Since $\mathcal{D}(G)$ is a Poisson Lie group, the above multiplication maps are Poisson maps. Since the projections are also Poisson maps, the left dressing action is a Poisson action. Finally, note that the dressing orbits are symplectic leaves (in particular, Poisson submanifolds). The case of the right dressing action can be considered similarly. □

EXAMPLE 6.1.4. Suppose that $G = SU(2)$. Denote by $DT(2)$ the group of dilations and translations of the Euclidean plane. It is isomorphic to the group of 2×2 matrices of the form

$$\begin{pmatrix} t & w \\ 0 & t^{-1} \end{pmatrix}, \qquad t > 0, \quad w \in \mathbb{C}.$$

Suppose that $SU(2)$ is equipped with the standard Poisson Lie group structure. We leave it as an easy exercise to show that the dual Poisson Lie group $SU(2)^*$ is isomorphic to $DT(2)$ as a Lie group.

Since the group $SU(2)$ is compact, by Exercise 3.4.11 there exists the global left dressing action of $DT(2)$ on $SU(2)$. By Proposition 6.1.3, the orbits of the dressing action are Poisson $DT(2)$-manifolds. By Proposition 3.4.10, the orbits of the dressing action are the symplectic leaves in $SU(2)$. These have been described in Example 5.2.4. On the zero-dimensional leaves the group $DT(2)$ acts trivially. The action of $DT(2)$ on the two-dimensional leaves (each isomorphic to the complex plane) is by dilations and translations.

The explicit formulas are as follows. Let $z \in \mathbb{C}$ be the coordinate on the symplectic leaf S_{-1} such that

$$S_{-1} = \left\{ \left(1 + |z|^2\right)^{-\frac{1}{2}} \begin{pmatrix} z & 1 \\ -1 & \bar{z} \end{pmatrix} \right\}.$$

Then the action of $DT(2)$ on S_{-1} is given by

$$\begin{pmatrix} t & w \\ 0 & t^{-1} \end{pmatrix} : z \mapsto t(tz + w).$$

The following theorem states the necessary and sufficient conditions for an action of a Poisson Lie group on a Poisson manifold to be Poisson.

THEOREM 6.1.5. *Let G be a connected Poisson Lie group, $\mathfrak{g}$ the Lie algebra of G, and M a G-manifold equipped with a Poisson manifold structure. Let π_G and π_M be the bivector fields corresponding to the Poisson manifold structures on G and M respectively. Then the following conditions are equivalent:*

(1) *The action $\sigma : G \times M \to M$ is a Poisson action, so that M is a Poisson G-manifold.*

(2) *For any $g \in G$ and $x \in M$,*

$$\pi_M(gx) = \left(\sigma'_g\right)_* \pi_M(x) + \left(\sigma''_x\right)_* \pi_G(g), \tag{6.1.1}$$

where $\sigma'_g : x \mapsto gx$ is a map $M \to M$, $\sigma''_x : g \mapsto gx$ a map $G \to M$, $\left(\sigma'_g\right)_*$ and $\left(\sigma''_x\right)_*$ the corresponding induced maps in the spaces of bivector fields.

(3) *For any $X \in \mathfrak{g}$, $f, g \in C^\infty(M)$,*

$$\zeta_X\{f,g\} - \{\zeta_X f, g\} - \{f, \zeta_X g\} = \langle [\theta_f, \theta_g], X \rangle, \tag{6.1.2}$$

where $\zeta : X \mapsto \zeta_X$ is the Lie algebra antihomomorphism from $\mathfrak{g}$ to the Lie algebra of vector fields on M corresponding to the infinitesimal action of $X \in \mathfrak{g}$, and θ_f is the $\mathfrak{g}^*$-valued function on M given by

$$\theta_f(x) = d_g f(gx)|_{g=e}, \quad x \in M. \tag{6.1.3}$$

PROOF. (1) $\iff$ (2). By definition, $\sigma : G \times M \to M$ is a Poisson map if and only if for any $\varphi, \psi \in C^\infty(M)$ and $g \in G, x \in P$, we have that

$$\{\varphi \circ \sigma, \psi \circ \sigma\}_{G \times M}(g, x) = \{\varphi, \psi\}_M(gx).$$

This is equivalent to

$$\{\varphi \circ \sigma''_x, \psi \circ \sigma''_x\}_G(g) + \{\varphi \circ \sigma'_g, \psi \circ \sigma'_g\}_M(x) = \{\varphi, \psi\}_M(gx).$$

But this relation is easily seen to be equivalent to (3.2.1).

(2) $\iff$ (3). By (6.1.1), we get that

$$(\sigma'_{g^{-1}})_* \pi_M(gx) = \pi_M(x) + (\sigma''_x)_* \left(l_{g^{-1}}\right)_* \pi_G(g).$$

Substituting $\exp tX$ for g and differentiating with respect to t at $t = 0$, we get that

$$\mathcal{L}_{\zeta_X} \pi_M = (\zeta \wedge \zeta)(d_e \pi_G)(X), \tag{6.1.4}$$

where $\mathcal{L}$ denotes the Lie derivative, and $d_e \pi_G : \mathfrak{g} \to \mathfrak{g} \wedge \mathfrak{g}$ is the inner derivative of π_G. Applying both sides of (6.1.4) to $df \wedge dg$, we get (6.1.2).

On the other hand, it is easy to see that (6.1.2) is equivalent to (6.1.4). Assuming now that (6.1.4) hold, we want to prove that

$$(\sigma'_{\exp(-tX)})_* \pi_M(\exp tX \, x) = \pi_M(x) + (\sigma''_x)_* \left(l_{\exp(-tX)}\right)_* \pi_G(\exp tX). \tag{6.1.5}$$

It is clear that (6.1.5) holds when $t = 0$. Differentiating both sides of (6.1.5) with respect to t, we get

$$\frac{d}{dt}\text{L.H.S.} = (\sigma'_{\exp(-tX)})_* (\mathcal{L}_{\zeta_X} \pi_M)(\exp tX \, x)$$

$$= (\sigma'_{\exp(-tX)})_*(\sigma''_{(\exp tX \, x)})_* (\mathcal{L}_X \pi_G)(e)$$

$$= (\sigma''_x)_* \text{Ad}_{\exp(-tX)} (\mathcal{L}_X \pi_G(e)),$$

$$\frac{d}{dt}\text{R.H.S.} = (\sigma''_x)_* \frac{d}{dt} \left(l_{\exp(-tX)}\right)_* \pi_G(\exp tX)$$

$$= (\sigma''_x)_* \text{Ad}_{\exp(-tX)} (\mathcal{L}_X \pi_G(e)).$$

Thus, we see that $\frac{d}{dt}$L.H.S. $= \frac{d}{dt}$R.H.S for any t. Therefore, (6.1.5) holds for any $x \in M$ and g in an open neighborhood of the unit element in G. Since G is connected, any open neighborhood of e in G generates G. But (6.1.5) is easily seen to be equivalent to (3.2.1). The theorem is proved. □

Now we will introduce the notion of moment map. It is well known for a Poisson action of a Lie group with the trivial Poisson structure on a symplectic manifold, but can be formulated to embrace the general case of a Poisson action.

Let G be a Poisson Lie group, $\mathfrak{g}$ the Lie bialgebra of G, G^* the Poisson Lie group dual to G, and M a left Poisson G-manifold (the case of right G-manifolds can be considered similarly). For any $X \in \mathfrak{g}$, let α_X be the left invariant differential 1-form on G^* with $\alpha_X(e) = X$. Let $X \mapsto \xi_X$ be the infinitesimal left action of $\mathfrak{g}$ on M, that is, a Lie algebra antihomomorphism (homomorphism in the case of a right action) from $\mathfrak{g}$ to the space of vector fields on M. Assume that the vector fields ξ_X give a global left action of G on M.

Recall that the bivector field π_M which defines the Poisson manifold structure on M gives rise to a map $\check{\pi}_M$ from the space of differential 1-forms on M to the space of vector fields on M.

DEFINITION 6.1.6. A map $J: M \to G^*$ is called a *moment map* if
$$\xi_X = \check{\pi}_M\left(J^*(\alpha_X)\right) \quad (\text{resp. } \zeta_X = -\check{\pi}_M\left(J^*(\beta_X)\right)).$$

EXERCISE 6.1.7. Give a definition of a moment map for a right Poisson G-manifold.

EXERCISE 6.1.8. A moment map does not necessarily exist for an arbitrary Poisson G-manifold. Check that the left action of G on itself by left multiplications does not admit a moment map.

REMARK 6.1.9. Perhaps, the simplest case when a moment map does exist is when we consider the left (or right) dressing action of G on $M = G^*$. It is easy to see that in this case the identity map $G^* \to G^*$ is a moment map.

For any symplectic manifold M, we have a short exact sequence
$$0 \to \mathbb{C} \xrightarrow{\iota} C^\infty(M) \xrightarrow{\xi} \mathcal{H}(M) \to 0,$$
where $C^\infty(M)$ is regarded as a Lie algebra with respect to the Poisson bracket, and $\mathcal{H}(M)$ is the Lie algebra of Hamiltonian vector fields on M. Here ι is the embedding of $\mathbb{C}$ as the center in $C^\infty(M)$, while ξ is given by $\varphi \mapsto \xi_\varphi = \{\varphi, \cdot\}$.

Suppose that G is a Poisson Lie group with the trivial Poisson structure. Let $\lambda(\xi)$ be the vector field corresponding to the infinitesimal action of $\xi \in \mathfrak{g}$ on M. That is, we have
$$\lambda(\xi)f(x) = \frac{d}{dt}\Big|_{t=0} f\left(e^{t\xi}x\right).$$

Since the Poisson structure on G is trivial, the vector fields $\lambda(\xi)$ preserve the symplectic form on M.

DEFINITION 6.1.10. For a Poisson Lie group G with trivial Poisson structure, a left symplectic G-manifold M is called *Hamiltonian G-manifold* if there exists a Lie algebra homomorphism $\mu: \mathfrak{g} \to C^\infty(M)$ such that $\lambda(\xi) = \iota(\mu(\xi))$ for any $\xi \in \mathfrak{g}$.

REMARK 6.1.11. For any left symplectic G-manifold M and $\xi \in \mathfrak{g}$, consider the differential 1-form $\beta(\xi)$ defined as the image of the vector field $\lambda(\xi)$ under the map $Vect(M) \to \Omega^1(M)$ induced by the canonical isomorphism $\check{\pi}^{-1}: TM \tilde{\to} T^*M$. By the Poincaré lemma, there is a function $f_{\xi,x}$ defined locally in a neighborhood of $x \in M$ such that $\beta(\xi) = df_{\xi,x}$ in a neighborhood of x.

Such collection of functions may or may not be extended to a global function f_ξ on M such that $\beta(\xi) = df_\xi$. The left Hamiltonian G-manifolds are exactly those for which such an extention exists.

EXAMPLE 6.1.12. For any left Hamiltonian G-manifold M, one can define a moment map as follows. Given a lift $\mu: \mathfrak{g} \to C^\infty(M)$, define a map $\check{\mu}: M \to \mathfrak{g}^*$ by $\check{\mu}(x)(a) = \mu(a)(x)$ for any $x \in M$ and $a \in \mathfrak{g}$. It is easy to see that $\check{\mu}$ satisfies Definition 6.1.6 of a moment map. Moreover, it G acts transitively on M, the image of $\check{\mu}$ is a coadjoint orbit in $\mathfrak{g}^*$.

EXERCISE 6.1.13. Let G be a compact Lie group. Prove that any left homogeneous Hamiltonian G-manifold is isomorphic to a finite cover of a coadjoint orbit in $\mathfrak{g}^*$.

EXAMPLE 6.1.14. If G is equipped with the trivial Poisson structure, then G^* is isomorphic to $\mathfrak{g}^*$ regarded as an abelian Lie group with the Kirillov–Kostant bracket, and the dressing action becomes the coadjoint one. Therefore, given a coadjoint orbit, the corresponding moment map is the embedding of the orbit into $\mathfrak{g}^*$.

EXAMPLE 6.1.15. Comparing Remark 6.1.9 with Example 6.1.12, we see that for any dressing G-orbit in G^*, there exists a moment map given by the embedding of the orbit into G^*.

For any left Hamiltonian G-manifold M such that the space of orbits $G \setminus M$ is a C^∞-manifold, there is a canonical full dual pair

(6.1.6) $$G \setminus M \xleftarrow{p} M \xrightarrow{\check{\mu}} \mathfrak{g}^*,$$

where $\check{\mu}$ is the moment map and p is the natural projection on the set of G-orbits. As an immediate consequence, we see that by Theorem 1.2.5, the blow-up $p \circ (\check{\mu})^{-1}(a)$ is a symplectic leaf in $G \setminus M$ for any $a \in \mathfrak{g}^*$. This fact is widely used in the technique of Hamiltonian reduction.

PROPOSITION 6.1.16. *The following formula defines a Poisson manifold structure on $\mathcal{D}(G)$:*

(6.1.7) $$\{f_1, f_2\} = \sum_\alpha (\partial_\alpha f_1 \partial^\alpha f_2 - \partial^\alpha f_1 \partial_\alpha f_2)$$
$$+ \sum_\alpha \left(\partial'_\alpha f_1 (\partial^\alpha)' f_2 - (\partial^\alpha)' f_1 \partial'_\alpha f_2\right).$$

The resulting Poisson manifold $\mathcal{D}_s(G)$ is symplectic in some neighborhood of the unit element of $\mathcal{D}(G)$. It is called the symplectic double *of G.*

PROOF. It was shown in the proof of Proposition 4.1.4 that the bracket (6.1.7) is, in fact, the sum of two Poisson brackets. Therefore, it defines a Poisson manifold structure on $\mathcal{D}(G)$. On the other hand, the linearization of the bracket (6.1.7) at the unit element in $\mathcal{D}(G)$ is nondegenerate, since the right- and left-invariant vector fields coincide at the unit element ($\partial_\alpha = \partial'_\alpha$ and $\partial^\alpha = (\partial^\alpha)'$). Therefore, it yields a symplectic form in a neighborhood of the unit element. □

PROPOSITION 6.1.17. *The Poisson manifold structure on the symplectic double $\mathcal{D}_s(G)$ induces Poisson manifold structures on $G^* \backslash \mathcal{D}_s(G)$ and $\mathcal{D}_s(G)/G^*$ such that the natural projections $p_1 : \mathcal{D}_s(G) \to G^* \backslash \mathcal{D}_s(G)$ and $p_2 : \mathcal{D}_s(G) \to \mathcal{D}_s(G)/G^*$ are Poisson maps. As Poisson manifolds, $\mathcal{D}_s(G)/G^*$ is isomorphic and $G^* \backslash \mathcal{D}_s(G)$ antiisomorphic to G in a neighborhood of the coset G^*.*

PROOF. Consider $\mathcal{D}_s(G)/G^*$. Let f_1 and f_2 be right G^*-invariant functions on $\mathcal{D}_s(G)$. By (6.1.7), we have that

$$\{f_1, f_2\}_{\mathcal{D}_s} = \sum_\alpha (\partial_\alpha f_1 \partial^\alpha f_2 - \partial_\alpha f_2 \partial^\alpha f_1) = \{f_1, f_2\}_{\mathcal{D}(G)}.$$

Thus, we see that not only the Poisson bracket of right G^*-invariant functions on $\mathcal{D}_s(G)$ is a right G^*-invariant function (which implies that $\mathcal{D}_s(G)/G^*$ is a Poisson manifold such that p_2 is a Poisson map), but it also coincides with the Poisson bracket of the same right G^*-invariant functions on $\mathcal{D}(G)$. By Proposition 3.4.3, $\mathcal{D}_s(G)/G^*$ is locally isomorphic to G as a Poisson manifold.

Finally, when we consider the case of $G^* \backslash \mathcal{D}_s(G)$, we similarly come to the conclusion that for any left G^*-invariant functions f_1 and f_2 on $\mathcal{D}_s(G)$,

$$\{f_1, f_2\}_{\mathcal{D}_s} = -\{f_1, f_2\}_{\mathcal{D}(G)}.$$

Therefore, $G^* \backslash \mathcal{D}_s(G)$ is locally antiisomorphic to G as a Poisson manifold. □

THEOREM 6.1.18. *The diagram*

(6.1.8) $$G \simeq G^* \backslash \mathcal{D}_s(G) \xleftarrow{p_1} \mathcal{D}_s(G) \xrightarrow{p_2} \mathcal{D}_s(G)/G^* \simeq G$$

is a full dual pair in a neighborhood of the unit element of $\mathcal{D}(G)$.

PROOF. We want to show that the algebras of left G^*-invariant and right G^*-invariant functions on $\mathcal{D}(G)$ centralize each other with respect to the Poisson bracket on $\mathcal{D}(G)$. Suppose that φ is a left G^*-invariant function, and ψ is a right G^*-invariant function.

Define the left gradient ∇_f and the right gradient ∇'_f of a function f on $\mathcal{D}(G)$ as $\mathcal{D}(\mathfrak{g})$-valued functions on $\mathcal{D}(G)$ given by

$$(\nabla_f(x), \alpha) = (\zeta(\alpha)f)(x) = \frac{d}{dt}\Big|_{t=0} f\left(e^{t\alpha}x\right),$$

$$(\nabla'_f(x), \alpha) = (\zeta'(\alpha)f)(x) = \frac{d}{dt}\Big|_{t=0} f\left(xe^{t\alpha}\right)$$

for any $\alpha \in \mathcal{D}(\mathfrak{g})$ and $x \in \mathcal{D}(G)$. Here $\zeta(\alpha)$ is the right invariant vector field corresponding to α, while $\zeta'(\alpha)$ is the corresponding left invariant vector field.

It follows from (6.1.7) that

(6.1.9) $$\{\varphi, \psi\}_{\mathcal{D}_s} = \sum_\alpha \left(\partial_\alpha \varphi \partial^\alpha \psi - \partial'_\alpha \varphi (\partial^\alpha)' \psi\right)$$
$$= (\nabla_\varphi, \nabla_\psi) - (\nabla'_\varphi, \nabla'_\psi).$$

Since φ is a left G^*-invariant function and ψ is a right G^*-invariant function, we get that $\nabla_\varphi(x), \nabla'_\psi(x) \in (\mathfrak{g}^*)^\perp = \mathfrak{g}$ for any $x \in \mathcal{D}(G)$. On the other hand, we have that for any $g_+ \in G$, $g_- \in G^*$ and $\alpha \in \mathfrak{g}^*$ the following holds:

$$(\nabla_\psi(g_+g_-), \alpha) = \frac{d}{dt}\Big|_{t=0} \psi\left(e^{t\alpha} g_+\right) = \frac{d}{dt}\Big|_{t=0} \psi\left(g_+^{e^{t\alpha}} \left(e^{t\alpha}\right)^{g_+}\right)$$

$$= \frac{d}{dt}\Big|_{t=0} \psi\left(g_+^{e^{t\alpha}}\right) = (\rho_l(\alpha)\psi)(g_+),$$

$$(\nabla'_\varphi(g_-g_+), \alpha) = \frac{d}{dt}\Big|_{t=0} \varphi\left(g_+ e^{t\alpha}\right) = \frac{d}{dt}\Big|_{t=0} \varphi\left(\left(e^{t\alpha}\right)^{g_+} g_+^{e^{t\alpha}}\right)$$

$$= \frac{d}{dt}\Big|_{t=0} \varphi\left(g_+^{e^{t\alpha}}\right) = (\rho_r(\alpha)\varphi)(g_+).$$

In particular, we get that

$$(6.1.10) \quad (\nabla_\varphi(g_+g_-), \nabla_\psi(g_+g_-)) = (\rho_l(\nabla_\varphi(g_+g_-))\psi)(g_+)$$
$$= \{\varphi(g_+g_-), \psi(g_+)\}_G$$
$$= \{\varphi(g_+g_-), \psi(g_+g_-)\}_G,$$

$$(6.1.11) \quad (\nabla'_\varphi(g_+g_-), \nabla'_\psi(g_+g_-)) = \left(\nabla'_\varphi\left(g_-^{g_+} g_+^{g_-}\right), \nabla'_\psi(g_+g_-)\right)$$
$$= \left(\rho_r\left(\nabla'_\psi(g_+g_-)\right) \varphi\right)\left(g_+^{g_-}\right)$$
$$= \left\{\varphi\left(g_+^{g_-}\right), \psi(g_+g_-)\right\}_G$$
$$= \{\varphi(g_+g_-), \psi(g_+g_-)\}_G.$$

It follows from (6.1.9), (6.1.11), and (6.1.12) that $\{\varphi, \psi\}_{\mathcal{D}_s} = 0$. This means that the fibers of p_1 and p_2 intersect orthogonally in a neighborhood of the unit element of $\mathcal{D}(G)$. The dimension arguments imply that they are, in fact, orthogonal complements to each other. We leave it as an exercise to the reader to show that both p_1 and p_2 are submersions locally. $\square$

REMARK 6.1.19. Under the conditions of Example 6.1.12, the diagram (6.1.6) is a global full dual pair of the form (6.1.8), where $M = T^*G$ with the standard symplectic structure.

REMARK 6.1.20. Note that Theorem 3.4.14 can also be deduced from Theorem 6.1.18. The new proof is a direct consequence of applying the result of Theorem 1.2.5 to the full dual pair (6.1.8).

6.2. Poisson homogeneous G-manifolds.

DEFINITION 6.2.1. Let G be a Poisson Lie group. A Poisson G-manifold M is called *homogeneous* if G acts on M transitively.

As is well known, any left homogeneous G-manifold M is of the form $M = G/H$, where H is a Lie subgroup of G.[2] However, if G is a Poisson Lie group, the list of homogeneous Poisson G-manifolds is much richer, since it may happen (and usually it does) that the same G-manifold possesses a family of nonequivalent Poisson G-manifold structures.

This is important for the purposes of the quantum orbit method, and in the second part of the book we will see how the different Poisson G-manifold structures can give rise to nonisomorphic irreducible $*$-representations of the quantized algebras.

DEFINITION 6.2.2. A Lie subgroup H of a Poisson Lie group G is called *Poisson Lie subgroup* if it is also a Poisson submanifold of G.

---

[2]Recall that we always consider only *closed* subgroups.

PROPOSITION 6.2.3. *Let H be a connected Lie subgroup of a Poisson Lie group G, and $\mathfrak{h}$ and $\mathfrak{g}$ the Lie algebras of H and G respectively. The following statements are equivalent:*

(1) *H is a Poisson Lie subgroup of G;*

(2) *$\mathfrak{h}^\perp$ is an ideal in $\mathfrak{g}^*$;*

(3) *H is invariant with respect to the right (therefore, also the left) dressing action of G^* on G if it is defined globally.*

PROOF. Assume that $\mathfrak{h}^\perp$ is an ideal in $\mathfrak{g}^*$. Since the cobracket φ in $\mathfrak{g}$ is dual to the Lie bracket in $\mathfrak{g}^*$, we get immediately that $\varphi(\mathfrak{h}) \subset \mathfrak{h} \wedge \mathfrak{h}$. This implies that $\mathfrak{h}$ is a Lie bialgebra. By Theorem 3.3.1, there exists a unique Poisson Lie group structure on G, as well as a unique Poisson Lie group structure on H corresponding to the cobracket φ. By the uniqeness, H is a Poisson Lie subgroup in G.

Assume that H is a Poisson Lie subgroup in G. Then the right (resp. left) dressing vector fields are tangent to H. Therefore, the statement (3) holds.

Finally, assume that H is invariant with respect to the right (or left) dressing action. This means that for any $h \in H \subset G$ and $g \in G^*$ from small neighborhoods of the unit element of $\mathcal{D}(G)$, $hg = g^h h^g$, where $h^g \in H \subset G$ and $g^h \in G^*$. It follows that for any $a \in \mathfrak{h} \subset \mathfrak{g}$ and $\alpha \in \mathfrak{g}^*$, $[\alpha, a] = b + \beta$ in $\mathcal{D}(\mathfrak{g})$, where $b \in \mathfrak{h}$ and $\beta \in \mathfrak{g}^*$. Given any $\gamma \in \mathfrak{h}^\perp$, we have $([\alpha, a], \gamma) = (b + \beta, \gamma) = 0$. But, on the other hand, we have $([\alpha, a], \gamma) = (a, [\alpha, \gamma])$. This implies that $[\alpha, \gamma] \in \mathfrak{h}^\perp$, so that $\mathfrak{h}^\perp$ is an ideal in $\mathfrak{g}^*$. □

PROPOSITION 6.2.4. *Let G be a Poisson Lie group, and H a Poisson Lie subgroup of G. Then there exists an induced Poisson manifold structure on G/H such that*

(1) *the natural projection $p : G \to G/H$ is a Poisson map;*

(2) *the natural action of G on G/H is a Poisson action (hence, G/H becomes a homogeneous Poisson G-manifold);*

(3) *the local right dressing action of G^* on G induces a local right action of G^* on G/H, which we will also call the right dressing action; the symplectic leaves in G/H are integral manifolds of the corresponding dressing vector fields; in particular, if the action can be defined globally, the symplectic leaves become the G^*-orbits in G/H.*

PROOF. Consider the bivector field π on G which corresponds to the Poisson Lie group structure. Define a bivector field on G/H by

(6.2.1) $$\pi_{G/H}(gH) = p_*\pi(g).$$

It follows from the "multiplicativity" (3.2.1) of π that

$$p_*\pi(gh) = p_*(l_g)_*\pi(h) + p_*(r_h)_*\pi(g).$$

Since $p : G \to G/H$ is a morphism of G-manifolds, we get

$$p_*(l_g)_*\pi(h) = (l_g)_*p_*\pi(h) = 0,$$

if $h \in H$. Thus,

$$p_*\pi(gh) = p_*(r_h)_*\pi(g) = p_*\pi(g),$$

so the bivector field $\pi_{G/H}$ is well defined.

The statement (1) now follows immediately from (6.2.1). The definition (6.2.1) together with (3.2.1) implies (6.1.1) for $M = G/H$, thus showing that G/H is a left Poisson G-space. Since H is invariant with respect to the right dressing action of

G^* on G, we get an induced action of G^* in G/H. We leave it as a simple exercise to check that this action is also Poisson. If it is defined globally, its orbits are the symplectic leaves in G/H, since the dressing fields $\rho_l(\alpha)$ ($\alpha \in \mathfrak{g}^*$) are tangent to the symplectic leaves in G. □

EXERCISE 6.2.5. Formulate and prove an analog of Proposition 6.2.4 for G/H.

In particular, the left (resp. right) action of G on itself by left (resp. right) multiplications is a Poisson action: the fact already known to us.

EXAMPLE 6.2.6. Let K be a compact simple Poisson Lie group. Recall that it has a maximal torus T (see Section 5) such that any point of T is a zero-dimensional symplectic leaf in K. Therefore, the flag manifold $X = K/T$ has a structure of a homogeneous Poisson K-manifold.

Since X is compact, the local dressing actions of G^* on X described in Proposition 6.2.4 can be defined globally. Hence, X is equipped also with a structure of a Poisson K^*-manifold, although not a homogeneous one. The K^*-orbits are the Schubert cells in this case.

EXERCISE 6.2.7. Let G be the complexification of K and $G_\mathbb{R}$ the corresponding real group. Recall that the flag manifold $X = K/T$ has a complex manifold structure, arising from the realization of X as a G-manifold $X \simeq G/B_+$. Show that the same Poisson manifold structure which is described in Example 6.2.6 makes X a homogeneous Poisson $G_\mathbb{R}$-manifold. Also, the global dressing action of $G_\mathbb{R}^*$ on X makes X a Poisson $G_\mathbb{R}^*$-manifold, although not a homogeneous one.

We have seen above how one can get examples of homogeneous Poisson G-manifolds as long as one finds a Poisson Lie subgroup of G. However, it is not necessary for a Lie subgroup H of G to be Poisson in order to have a Poisson G-manifold structure on G/H. By Proposition 6.2.3, H is a Poisson Lie subgroup of G if and only if $\mathfrak{h}^\perp$ is an ideal in $\mathfrak{g}^*$.

DEFINITION 6.2.8. Let $\mathfrak{g}$ be a Lie bialgebra, and $\mathfrak{h}$ a Lie subalgebra of $\mathfrak{g}$.
(1) We call $\mathfrak{h}$ *Lie sub-bialgebra* of $\mathfrak{g}$ if $\mathfrak{h}^\perp$ is an ideal in $\mathfrak{g}^*$.
(2) We call $\mathfrak{h}$ *Lie bi-subalgebra* of $\mathfrak{g}$ if $\mathfrak{h}^\perp$ is a Lie subalgebra in $\mathfrak{g}^*$.

EXAMPLE 6.2.9. Consider a simple complex Lie algebra $\mathfrak{g}$ equipped with the standard Lie bialgebra structure (cf. Example 2.1.9). Then $\mathfrak{b}_\pm$ are Lie sub-bialgebras of $\mathfrak{g}$, whereas $\mathfrak{n}_\pm$ are Lie bi-subalgebras of $\mathfrak{g}$.

PROPOSITION 6.2.10. *Let G be a Poisson Lie group, and $\mathfrak{g}$ the Lie bialgebra of G. Suppose that H is a Lie subgroup of G such that the Lie algebra $\mathfrak{h}$ of H is a Lie bi-subalgebra of $\mathfrak{g}$. Then there exists an induced Poisson manifold structure on G/H such that*
(1) *the natural projection $p: G \to G/H$ is a Poisson map;*
(2) *the natural action of G on G/H is a Poisson action (hence, G/H becomes a homogeneous Poisson G-manifold).*

PROOF. Let ζ_X be the vector field on G corresponding to the infinitesimal right action of $X \in \mathfrak{g}$. Let φ and ψ be right H-invariant smooth functions on G. Then for any $X \in \mathfrak{h}$, $\zeta_X \varphi = \zeta_X \psi = 0$. Moreover, we have that $\theta_\varphi, \theta_\psi \in \mathfrak{h}^\perp \subset \mathfrak{g}^*$, where θ_φ and θ_ψ are defined by (6.1.3). By (6.1.2), we get that

$$\zeta_X \{\varphi, \psi\} = \langle [\theta_\varphi, \theta_\psi], X \rangle = 0.$$

It follows that $\{\varphi, \psi\}$ is also a right H-invariant function. Since the right H-invariant functions form a Lie subalgebra in $C^\infty(G)$ with respect to the Poisson bracket, we conclude that the quotient space G/H is a Poisson manifold, and the natural projection $G \to G/H$ is a Poisson map. This immediately implies that the natural action of G on G/H is a Poisson action. $\square$

EXERCISE 6.2.11. State and prove the similar proposition for $H\backslash G$.

EXAMPLE 6.2.12. Suppose that $G = SL_2(\mathbb{C})_\mathbb{R}$, that is, $SL_2(\mathbb{C})$ regarded as a real Lie group. Consider the standard Poisson Lie group structure on G. Recall that the corresponding Lie bialgebra structure on $\mathfrak{g} = \mathfrak{sl}(2,\mathbb{C})_\mathbb{R}$ is given in Example 2.1.7. Let ξ be an arbitrary nonzero element of $\mathfrak{g}$ which is not proportional to H. This means that $\xi = aX^+ + bX^- + cH$ with $|a|^2 + |b|^2 > 0$. We see that $\varphi(\xi) = \xi \wedge H$. Therefore, the Borel subalgebra $\mathfrak{b}_\xi$ that is generated by H and ξ is a Lie sub-bialgebra of $\mathfrak{g}$. (In particular, it does not depend on c.)

Let B_ξ be the Lie subgroup of $SL_2(\mathbb{C})_\mathbb{R}$ whose Lie algebra is $\mathfrak{b}_\xi$. By Proposition 6.2.10, the flag manifold $X_\xi = G/B_\xi \simeq \mathbb{C}P^1$ is equipped with the structure of a homogeneous Poisson $SL_2(\mathbb{C})_\mathbb{R}$-manifold.

EXERCISE 6.2.13. Suppose that $G = SL_2(\mathbb{C})_\mathbb{R}$. Consider the projective space $\mathbb{C}P^1$ with homogeneous coordinates (a,b). Introduce the coordinate $\gamma = \frac{a}{b}$ in the open chart $b \neq 0$ of $\mathbb{C}P^1$. Assume that $\gamma = \infty$ when $b = 0$.

Show that the full list of the equivalence classes of Poisson G-manifold structures on the flag manifold $X = G/B$, where B is a Borel subgroup of $G = SL_2(\mathbb{C})_\mathbb{R}$, is given by the homogeneous Poisson $SL_2(\mathbb{C})_\mathbb{R}$-manifolds X_ξ described below, where $\xi = aX_+ + bX_-$, parameterized by $(a,b) \in \mathbb{C}P^1$, subject to the equivalences $(a,b) \mapsto (-a,b)$ and $(a,b) \mapsto (b,a)$ (or $\gamma \mapsto -\gamma$ and $\gamma \mapsto \gamma^{-1}$ respectively):

(1) $\gamma = 0$ (or, equivalently, $\gamma = \infty$); this structure is usually called *standard* because in this case $\mathfrak{b}_\xi$ is a Lie bi-subalgebra of $\mathfrak{g}$;
(2) $\gamma \in \mathbb{C} \setminus \{0\}$; these structures are usually called *nonstandard* because $\mathfrak{b}_\xi$ is not a Lie bi-subalgebra but just a Lie sub-bialgebra of $\mathfrak{g}$.

EXAMPLE 6.2.14. If in Exercise 6.2.13 we take $\gamma \in \mathbb{R}$, then the flag manifold $X_\xi \simeq \mathbb{C}P^1 \simeq S^2$ is also a homogeneous Poisson $SU(2)$-manifold.

However, it is not true in general that any left Poisson homogeneous G-manifold is of the form G/H, where H is a Lie subgroup of G whose Lie algebra $\mathfrak{h}$ is a Lie bi-subalgebra of $\mathfrak{g}$. In fact, the left Poisson homogeneous G-manifolds M obtained in this way are precisely those that contain a point $x \in M$ such that the Poisson bivector field π_M vanishes at x (consider the image of the unit element in G with respect to the natural projection $G \to M = G/H$). This condition fails already if $G = \mathbb{R}^{2n}$ with the trivial Poisson structure and $M = \mathbb{R}^{2n}$ with the standard Poisson bracket

$$\{f, g\} = \sum_{i=1}^{n} \left(\frac{\partial f}{\partial x_i} \frac{\partial g}{\partial x_{n+i}} - \frac{\partial f}{\partial x_{n+i}} \frac{\partial g}{\partial x_n} \right),$$

which defines a symplectic structure on $\mathbb{R}^{2n}$. Below we give Drinfeld's classification of the left Poisson homogeneous G-manifolds.

Let G be a Poisson Lie group. Consider the adjoint action of its Lie bialgebra $\mathfrak{g}$ on the double Lie bialgebra $\mathcal{D}(\mathfrak{g})$ and lift it to the action of the group.

6. POISSON G-MANIFOLDS

Let M be a left homogeneous G-space, and π_M a bivector field defining a Poisson manifold structure on M. For any $x \in M$, we have that $\pi_M(x) \in T_xM \wedge T_xM$, where $T_xM \simeq \mathfrak{g}/\mathfrak{h}_x$. Here $\mathfrak{h}_x$ is the Lie algebra of the stabilizer of x.

PROPOSITION 6.2.15. *For any $v \in (\mathfrak{g}/\mathfrak{h}_x) \wedge (\mathfrak{g}/\mathfrak{h}_x)$, the linear space*
$$L = \{a + \alpha \mid a \in \mathfrak{g}, \alpha \in \mathfrak{h}_x^\perp \subset \mathfrak{g}^*, (\alpha \otimes \mathrm{id})(v) = \bar{a}\},$$
where $\bar{a} \in \mathfrak{g}/\mathfrak{h}_x$ is the image of a under the natural projection $\mathfrak{g} \to \mathfrak{g}/\mathfrak{h}_x$, is a Largangian (that is, maximal isotropic) subspace in $\mathcal{D}(\mathfrak{g})$.

PROOF. Indeed, for any $a + \alpha, b + \beta \in L$, we have $\langle a, b \rangle = \langle \alpha, \beta \rangle = 0$ due to the fact that both $\mathfrak{g}$ and $\mathfrak{g}^*$ are isotropic in $\mathcal{D}(\mathfrak{g})$. On the other hand,
$$\langle a, \beta \rangle + \langle b, \alpha \rangle = (\alpha \otimes \beta + \beta \otimes \alpha)(v) = 0.$$
Therefore, L is an isotropic subspace in $\mathcal{D}(\mathfrak{g})$.

To show that L is a maximal isotropic subspace, note that the linear map $a + \alpha \mapsto a$ is a surjective map from L to $\mathfrak{g}$. By the dimension argument, we get that L is maximal. The proposition is proved. □

THEOREM 6.2.16. *For any $x \in M$, let L_x be the Lagrangian subspace in $\mathcal{D}(\mathfrak{g})$ corresponding to $v = \pi_M(x)$. Then the following statements hold:*

(1) *L_x is a Lie subalgebra in $\mathcal{D}(\mathfrak{g})$ for any $x \in M$.*

(2) *$L_{gx} = g(L_x)$ for any $g \in G$ and $x \in X$, where $g(L_x)$ denotes the adjoint action of G in $\mathcal{D}(\mathfrak{g})$.*

(3) *There is a bijection between Poisson G-manifold structures on a left homogeneous G-manifold M and G-equivariant maps from M to the set Λ of all Lagrangian Lie subalgebras in $\mathcal{D}(\mathfrak{g})$ such that if $x \in M$ and L_x is the corresponding Lagrangian subalgebra in $\mathcal{D}(\mathfrak{g})$, then $L_x \cap \mathfrak{g} = \mathfrak{h}_x$ is the Lie algebra of the stabilizer of x.*

PROOF. We start with proving (2) and (3), and then prove (1). To prove (2), note first that for any $g \in G$, $g(\alpha) \in \mathfrak{h}_{gx}^\perp$ if and only if $\alpha \in \mathfrak{h}_x^\perp$. Thus, we only have to show that
$$(g(\alpha) \otimes \mathrm{id})(\pi_M(gx)) = \overline{g(a)},$$
provided that $(\alpha \otimes \mathrm{id})(\pi_m(x)) = \bar{a}$. This is equivalent to
$$\langle \pi_M(gx), g(\alpha) \otimes \beta \rangle = \langle g(\alpha), \beta \rangle$$
for any $\beta \in \mathfrak{h}_{gx}^\perp$. It is clear that any such β is of the form $\beta = g(\gamma)$, where $\gamma \in \mathfrak{h}_x^\perp$. We have the equality
$$\langle \pi_M(gx), g(\alpha) \otimes g(\gamma) \rangle = \langle g(\alpha), g(\gamma) \rangle,$$
which is equivalent to $(\alpha \otimes \mathrm{id})(\pi_M(x)) = \bar{a}$. In particular, it suffices to prove that L_x is a Lie subalgebra for a single point $x \in M$ to have it proved for any point in M.

For any $x \in M$, there is a bijection between $T_xM \wedge T_xM$ and the set of all Lagrangian subspaces $L \subset \mathcal{D}(\mathfrak{g})$ such that $L \cap \mathfrak{g} = \mathfrak{h}_x$. Therefore, there is a bijection between the bivector fields on M and the smooth maps $x \mapsto L_x$ from M to the manifold of all Lagrangian subspaces in $\mathcal{D}(\mathfrak{g})$.

Let ξ be the bivector field compatible with the G-action on M. Consider the corresponding map $x \mapsto L_x$. It is G-equivariant. In particular, it follows that $g(L_x) = L_{gx} = L_x$ for any $g \in H_x$, where H_x is the stabilizer of $x \in M$. Therefore,
$$[\mathfrak{h}_x, L_x] \subset L_x. \tag{6.2.2}$$

We want to show that the bracket corresponding to ξ satisfies the Jacobi identity (hence defines a Poisson manifold structure on M compatible with the G-action) if and only if there exists $x \in M$ such that L_x is a Lie subalgebra in $\mathcal{D}(\mathfrak{g})$ (hence, $L_x \in \Lambda$ for any $x \in M$). This will also prove the third statement of the theorem.

Fix $x \in M$ and identify M with G/H_x. It is easy to see that ξ defines a Poisson bracket if and only if
$$\xi(gx) = \sigma'_g \xi(x) + \eta(g),$$
where $\eta(g) : G \to \mathfrak{g} \wedge \mathfrak{g}$ is the 1-cocycle corresponding to the Poisson Lie group structure on G given by (3.2.2). It follows that for any $f_1, f_2 \in C^\infty(M)$,
$$\{\phi,\psi\} = \sum_{\alpha,\beta} \left(\eta^{\alpha\beta}(g)(\partial_\alpha f_2)(g)(\partial_\beta f_2)(g) + \zeta^{\alpha\beta}(\partial'_\alpha f_1)(g)(\partial'_\beta f_2)(g) \right),$$
where ∂_α (resp. ∂'_α) is the right (resp. left) invariant vector field on G whose value at the unit element lies in a basis $\{I_\alpha\}$ of $\mathfrak{g}$. Also,
$$\eta(g) = \sum_{\alpha,\beta} \eta^{\alpha\beta}(g) I_\alpha \wedge I_\beta,$$
$$\zeta = \sum_{\alpha,\beta} \zeta^{\alpha\beta} I_\alpha \wedge I_\beta,$$
where $\zeta \in \mathfrak{g} \wedge \mathfrak{g}$ is any tensor whose image in $\mathfrak{g}/\mathfrak{h}_x \wedge \mathfrak{g}/\mathfrak{h}_x$ is equal to $\xi(x)$. One can show by a straightforward computation that for any $f_1, f_2, f_3 \in C^\infty(M)$, we have that

(6.2.3)
$$\{\{f_1, f_2\}, f_3\} + \{\{f_2, f_3\}, f_1\} + \{\{f_3, f_1\}, f_2\} = s^{\alpha\beta\gamma}(\partial'_\alpha f_1)(\partial'_\beta f_2)(\partial'_\gamma f_3),$$
where $s^{\alpha\beta\gamma}$ are the coordinates of the tensor
$$s = \left(\varphi^{12} + \varphi^{13} + \varphi^{23}\right)(r) - [r^{12}, r^{13}] - [r^{12}, r^{23}] - [r^{13}, r^{23}],$$
where $\varphi^{12}(r) = (\varphi \otimes \mathrm{id})(r)$, $\varphi^{23}(r) = (\mathrm{id} \otimes \varphi)(r)$ and $\varphi^{13}(r) = \sum_k a_k \otimes 1 \otimes b_k$ if $\varphi(r) = \sum_k a_k \otimes b_k$. Here $\varphi : \mathfrak{g} \to \mathfrak{g} \wedge \mathfrak{g}$ stands for the cobracket.

The right-hand side of (6.2.3) vanishes if and only if the image of s in $\bigwedge^3(\mathfrak{g}/\mathfrak{h}_x)$ vanishes. By (6.2.2), the latter happens if and only if L_x is a Lie subalgebra in $\mathcal{D}(\mathfrak{g})$. This completes the proof of the theorem. $\square$

COROLLARY 6.2.17. *There exists a bijection between the set of isomorphism classes of left Poisson homogeneous G-manifolds and the set of G-equivalence classes of pairs (L, H), where L is a Lagrangian subalgebra in $\mathcal{D}(\mathfrak{g})$ and H is a closed subgroup in $G_L = \{g \in G \mid g(L) = L\}$ such that $\mathrm{Lie}\, H = L \cap \mathfrak{g}$. Here two pairs (L_1, H_1) and (L_2, H_2) are called G-equivalent if there is $g \in G$ such that $L_2 = g(L_1)$ and $H_2 = g H_1 g^{-1}$.*

PROOF. Given a Poisson G-manifold M, take a point $x \in M$. Then the Lagrangian subalgebra $L = L_x$ in $\mathcal{D}(G)$ and the stabilizer $H = H_x$ of x define the corresponding pair. Let us show that isomorphic Poisson homogeneous G-manifolds define G-equivalent pairs.

It is clear that a choice of a Poisson homogeneous G-manifold isomorphic to M is equivalent to the choice of a point in M. Suppose that given points x_i, we get pairs (L_i, H_i), where $i = 1, 2$. Since G acts on M transitively, there exists $g \in G$ such that $x_2 = g x_1$. Then $H_2 = H_{x_2}$ (the stabilizer of x_2 is equal to $g H_1 g^{-1}$, where $H_1 = H_{x_1}$ is the stabilizer of x_1). Also, by the second statement of Proposition

6.2.16, $L_2 = L_{x_2}$ is equal to $g(L_1) = g(L_{x_1})$. In particular, equivalent left Poisson homogeneous G-manifolds correspond to G-equivalent pairs.

Vice versa, given a pair (L, H) as described above, consider the G-manifold $M = G/H$. Then define a G-equivariant map $M \to \Lambda$ by $gH \mapsto g(L)$. By Theorem 6.2.16, it corresponds to a structure of left Poisson homogeneous G-manifold on M. The choice of G-equivalent pairs (L_1, H_1) and (L_2, H_2) simply corresponds to the choice of the points $x_1, x_2 \in M$ such that $H_i = H_{x_i}$. It follows that the G-equivalent pairs define isomorphic Poisson homogeneous G-manifolds.

It can be readily checked that the constructed maps between the set of isomorphism classes of left Poisson homogeneous G-manifolds and the set of G-equivalence classes of the pairs (L, H) are inverse to each other. Thus, we get a bijection. □

7. Historical remarks

The material of Section 1 is standard and well known. In presenting it we basically follow [**Wei83**]. The notions of Lie bialgebra, Poisson Lie group, and Manin triple discussed in Sections 2 and 3 are due to Drinfeld. Most of the definitions and results are taken from [**Dri87**]. The useful Theorem 6.1.18 is due to Semenov-Tian-Shansky (cf. [**STS85**]). According to a reference in [**STS85**], it was proved also by Drinfeld (unpublished). The notion of dressing action (also called dressing transformation in literature) goes back to a number of papers on nonlinear integrable systems. For Poisson Lie groups it was introduced in [**STS85**], where several important applications were considered. Many intesting results on the dressing action were obtained by Lu and Weinstein (see, for instance, [**Lu91, LW90**]).

The material of Section 4 is presented according to [**BD84, Dri87**], where it was first introduced. The classification of compact simple Poisson Lie groups given in Section 5 was first obtained in [**Soi90, Soi89**]. The results about the structure of the symplectic leaves play a role similar to that of the coadjoint orbits in the representation theory.

The symplectic leaves in a compact simple Poisson Lie group K were described in the following papers: (1) in [**SV86, SV88**] for $K = SU(2)$; (2) in [**Soi89**] for $K = SU(n)$; (3) for $K = K(1,0)$, the relation to the Schubert cells and the Bruhat decomposition was observed independently in [**LW90**] and [**Soi90**]; the general result of Theorem 5.2.5 was obtained in [**Soi90**]; (4) the case $K = K(a,u)$, $u \neq 0$ is considered in [**LS91b**]. The symplectic leaves in a simple complex Poisson Lie group were described in [**HL93**].

The main results of Section 6 are largely due to Drinfeld. We can mention the definitions of Lie sub-bialgebra and Lie bi-subalgebra and the classification of Poisson G-spaces in Theorem 6.2.16 and Corollary 6.2.17 as examples. We are grateful to E. Karolinsky for sending us the original Drinfeld's proof of Theorem 6.2.16, which is shorter than ours. Some of the results are taken from [**STS85**] (Theorem 6.1.5, for instance); Proposition 6.2.4 is taken from [**LW90**]. The generalized concept of a moment map for any Poisson G-manifold was introduced in [**Lu91**].

CHAPTER 2

Quantized Universal Enveloping Algebras

We start with the problem of quantization of Lie bilagebras. Our main example is the quantized universal enveloping algebra of a complex simple Lie algebra. We discuss the structure of its center and the R-matrix. We give a short account of braided monoidal categories (strict case only). Then we consider an example given by the category of representations of the quantized universal enveloping algebra. Our presentation includes the case of a fixed quantization parameter and the so-called twisted case.

1. Quantization of Lie bialgebras

In this section we discuss the problem of quantization of Lie bialgebras. Our main example is the quantization of the standard Lie bialgebra structure on a complex simple Lie algebra.

1.1. Definition of quantization. We start with a (finite-dimensional) Lie bialgebra $(\mathfrak{g}, \varphi)$ over a field $\mathbf{k}$ of characteristic zero. Recall (see Examples 2.2.2, 2.2.10, and Definition 2.2.9 of Chapter 1) that the universal enveloping algebra $U\mathfrak{g}$ equipped with the extension (2.2.9) of a cobracket φ on $\mathfrak{g}$ becomes a co-Poisson Hopf algebra.

For any $\mathbf{k}[[h]]$-module V, we consider the topology defined by the system $h^k V$ ($k = 1, 2, \ldots$) of open neighborhoods of zero. We denote by $\hat{V}$ the completion of V in the h-adic topology. For any two $\mathbf{k}[[h]]$-modules V_1 and V_2, we define a $\mathbf{k}[[h]]$-module $V_1 \hat{\otimes} V_2$ as the completion $\widehat{V_1 \otimes V_2}$ of the tensor product, that is,

$$V_1 \hat{\otimes} V_2 = \varprojlim{}_n V_1/h^n V_1 \otimes V_2/h^n V_2.$$

REMARK 1.1.1. When referring to A as a Hopf algebra over $\mathbf{k}[[h]]$, we mean that A is a topological (with respect to the h-adic topology) algebra over $\mathbf{k}[[h]]$ equipped with a comultiplication $\Delta : A \to A \hat{\otimes} A$, an antipode $S : A \to A$, and a counit $\varepsilon : A \to \mathbf{k}[[h]]$ which are continuous in the h-adic topology and satisfy all the conditions of Definition 2.2.1 of Chapter 1, with all the tensor products replaced by tensor products $\hat{\otimes}$ completed in the h-adic topology.

DEFINITION 1.1.2. Let U_0 be a co-Poisson Hopf algebra over $\mathbf{k}$, and $\delta : U_0 \to U_0 \otimes U_0$ the corresponding cobracket. *Quantization* (or *formal quantization*) of U_0 is a Hopf algebra U over $\mathbf{k}[[h]]$ satisfying the following conditions:

(1) U is a topologically free $\mathbf{k}[[h]]$-module with respect to the h-adic topology, i.e., U is isomorphic to $U_0[[h]]$ as a topological $\mathbf{k}[[h]]$-module;
(2) U is complete in the h-adic topology;
(3) U/hU is isomorphic to U_0 as a Hopf algebra;

(4) $h^{-1}(\Delta(a) - \Delta'(a)) \bmod h = \delta(a \bmod h)$ for any $a \in U$, where Δ is the comultiplication in U, $\Delta' = \sigma\Delta$ is the inverse comultiplication, and $\sigma : a \otimes b \mapsto b \otimes a$ is the permutation in $U \otimes U$.

DEFINITION 1.1.3. Let $(\mathfrak{g}, \varphi)$ be a Lie bialgebra. Then a quantization U of the corresponding co-Poisson Hopf algebra $U_0 = U\mathfrak{g}$ is called *quantization* of $(\mathfrak{g}, \varphi)$. Equivalently, we will say that $(\mathfrak{g}, \varphi)$ is the *quasi-classical analog* (or the *quasi-classical limit*) of U.

A Hopf algebra over $\mathbf{k}[[h]]$ which is a quantization of some Lie bialgebra $(\mathfrak{g}, \varphi)$ will be called also a *quantized universal enveloping algebra* of $(\mathfrak{g}, \varphi)$ or simply a *QUE-algebra*.

EXAMPLE 1.1.4. Consider the Lie bialgebra $\mathfrak{g}$ over $\mathbf{k}$ with the trivial cobracket (see Example 2.1.7 of Chapter 1). Then the Hopf algebra $U\mathfrak{g}[[h]]$ is easily seen to be a quantization of $\mathfrak{g}$.

EXAMPLE 1.1.5. Consider the Lie bialgebra of Example 2.1.6 of Chapter 1, that is, the Lie bialgebra $\mathfrak{g}^*$ dual to the Lie bialgebra $\mathfrak{g}$ given in Example 2.1.7 of Chapter 1 (and quantized in Example 1.1.4). (Recall that $\mathfrak{g}^*$ is abelian as a Lie algebra and the cobracket φ on $\mathfrak{g}^*$ is dual to the Lie bracket in $\mathfrak{g}$.) Then the Hopf algebra $\mathbf{k}[[h]]\hat{\otimes} U\mathfrak{g}^*$ is a quantization of the Lie algebra $\mathfrak{g}^*$.

1.2. The quantization of complex simple Lie algebras. Suppose that $A = (a_{ij})_{i,j=1}^n$ is a symmetrizable Cartan matrix, and let $\{\alpha_i\}_{i=1}^n$, $\{h_i\}_{i=1}^n$ be systems of simple roots and coroots respectively such that $\langle h_i, \alpha_j \rangle = a_{ij}$ (what is called a *realization* of A in [**Kac90**]).

We denote by $\mathfrak{g}(A)$ the complex simple Lie algebra defined by the Cartan matrix A. Assume also that we have a fixed invariant scalar product on $\mathfrak{h}^* = \bigoplus_{i=1}^n \mathbb{C}\alpha_i$ such that $a_{ij} = 2\frac{(\alpha_i, \alpha_j)}{(\alpha_j, \alpha_j)}$.

Consider the Hopf algebra $U_h\mathfrak{g} = U_h\mathfrak{g}(A)$ over $\mathbb{C}[[h]]$ complete in h-adic topology and topologically generated by X_i^+, X_i^-, H_i ($i = 1, \ldots, n$) subject to the relations

(1.2.1) $$[H_i, H_j] = 0, \quad [H_i, X_j^\pm] = \pm(\alpha_i, \alpha_j) X_j^\pm,$$

(1.2.2) $$X_i^+ X_j^- - X_j^- X_i^+ = \delta_{ij} \frac{\sinh\left(\frac{h}{2} H_i\right)}{\sinh \frac{h}{2}},$$

(1.2.3) $$\sum_{k=0}^{1-a_{ij}} (-1)^k \binom{1-a_{ij}}{k}_{q_i} (X_i^\pm)^k X_j^\pm (X_i^\pm)^{1-a_{ij}-k} = 0 \quad \text{for } i \neq j,$$

where $q_i = q^{\frac{(\alpha_i, \alpha_i)}{2}}$ (so that $q_i^{a_{ij}} = q_j^{a_{ji}}$), $q = e^{\frac{h}{2}}$. Here $\binom{m}{n}_t = \frac{(t;t)_m}{(t;t)_n (t;t)_{m-n}}$ is the so-called *t-binomial coefficient* (or *Gauss polynomial*), where $(a;t)_k = (1-a)(1-at)\cdots(1-at^{k-1})$.

PROPOSITION 1.2.1. *There exists a Hopf algebra structure on $U_h\mathfrak{g}(A)$ with the comultiplication Δ, the antipode S, and the counit ε (continuous in the h-adic topology) given by*

(1.2.4) $\Delta(H_i) = H_i \otimes 1 + 1 \otimes H_i, \qquad \Delta(X_i^\pm) = q^{-\frac{H_i}{2}} \otimes X_i^\pm + X_i^\pm \otimes q^{\frac{H_i}{2}},$

(1.2.5) $\qquad\qquad S(H_i) = -H_i, \qquad S(X_i^\pm) = -q_i^{\pm 1} X_i^\pm,$

(1.2.6) $\qquad\qquad \varepsilon(H_i) = 0, \qquad \varepsilon(X_i^\pm) = 0.$

1. QUANTIZATION OF LIE BIALGEBRAS

PROOF. It is a straightforward computation to check that Δ, S, and ε are compatible with (1.2.1)–(1.2.3). Note that (1.2.5)–(1.2.6) actually follow from (1.2.4). □

EXERCISE 1.2.2. Let $\mathfrak{g} = \mathfrak{g}(A)$ be the finite-dimensional simple complex Lie algebra defined by the Cartan matrix A. Consider the standard Lie bialgebra structure on $\mathfrak{g}$ described in Example 2.1.9 and Exercise 2.3.7 of Chapter 1. Then the topological Hopf algebra $U_h\mathfrak{g}$ over $\mathbb{C}[[h]]$ defined above is a quantization of this Lie bialgebra.

Properties (2)–(4) in Definition 1.1.2 can be checked by straightforward computation. The only nontrivial part of the above exercise is to prove that $U_h\mathfrak{g}$ is a topologically free $\mathbb{C}[[h]]$-module.

DEFINITION 1.2.3. Let A be a Hopf algebra. The action of A on itself given by

$$\text{(1.2.7)} \qquad \text{ad}(a) : b \mapsto \sum_k a_k^{(1)} b S\left(a_k^{(2)}\right), \text{ where } \Delta(a) = \sum_k a_k^{(1)} \otimes a_k^{(2)},$$

is called the *adjoint action*. If A is a QUE-algebra, we will refer to it as the *quantum left adjoint action*.

REMARK 1.2.4. Note that the so-called quantum Serre relations (1.2.3) can be rewritten in the following more convenient (and more resembling the classical case) form:

$$\text{(1.2.8)} \qquad \text{ad}_q^{1-a_{ij}}(E_i) E_j = 0, \quad \text{ad}_q^{1-a_{ij}}(F_i) F_j = 0 \quad \text{for } i \neq j,$$

using another set of generators

$$\text{(1.2.9)} \qquad E_i = X_i^+ q^{-\frac{H_i}{2}}, \quad F_i = X_i^- q^{\frac{H_i}{2}}.$$

Here ad_q stands for the quantum adjoint action of $U_h\mathfrak{g}(A)$, which we also denote sometimes by ad. It is easy to see that the following formulas hold:

$$\text{(1.2.10)} \qquad \Delta(E_i) = E_i \otimes 1 + q^{-H_i} \otimes E_i,$$
$$\text{(1.2.11)} \qquad \Delta(F_i) = 1 \otimes F_i + F_i \otimes q^{H_i},$$
$$\text{(1.2.12)} \qquad S(E_i) = -q^{H_i} E_i, \qquad S(F_i) = -F_i q^{-H_i},$$
$$\text{(1.2.13)} \qquad \varepsilon(E_i) = 0, \qquad \varepsilon(F_i) = 0.$$

REMARK 1.2.5. Consider the Hopf subalgebra of $U_h\mathfrak{g}(A)$ generated by all H_i and X_i^+ ($i = 1, \ldots, n$). It is easy to see that it is a quantization of the Borel subbialgebra $\mathfrak{b}_+$ of $\mathfrak{g}(A)$. We denote it by $U_h\mathfrak{b}_+$. We define $U_h\mathfrak{b}_-$ similarly. For a fixed i we denote by $U_h\mathfrak{sl}(2)_i$ the Hopf subalgebra of $U_h\mathfrak{g}(A)$ generated by X_i^+, X_i^-, and H_i.

Now we formulate the following important result about the structure of the algebra $U_h\mathfrak{g}(A)$ in the case $\dim \mathfrak{g}(A) < \infty$. Namely, the quantized and the classical universal enveloping algebras are isomorphic as algebras. What is actually deformed is the coalgebra structure. In what follows we assume that $\mathfrak{g} = \mathfrak{g}(A)$, $\mathfrak{h}$ is the subspace generated by H_i ($i = 1, \ldots, n$), and $U_h\mathfrak{h}$ the Hopf subalgebra generated by $\mathfrak{h}$.

THEOREM 1.2.6. *Suppose that $\mathfrak{g}$ is a finite-dimensional complex simple Lie algebra with the standard Lie bialgebra structure (see Example 2.1.8 of Chapter 1). Then there exists an algebra isomorphism $U_h\mathfrak{g} \xrightarrow{\sim} (U\mathfrak{g})[[h]]$ which is the identity modulo $h\mathbb{C}[[h]]$. Moreover, this isomorphism can be chosen so that it is the identity map on $\mathfrak{h}$. The center of $U_h\mathfrak{g}$ is canonically isomorphic to $Z[[h]]$, where Z is the center of $U\mathfrak{g}$.*

PROOF. Formal deformations of an associative algebra A are parameterized by the Hochschild cohomology $H^2(A, A)$ (cf. [**GS88**]). It is easy to see that for $A = U\mathfrak{g}$, we have $H^2(A, A) = H^2(\mathfrak{g}, U\mathfrak{g})$, where $U\mathfrak{g}$ is a $\mathfrak{g}$-module via the adjoint action. It is well known (see [**Fuk86**] or [**Gui80**]) that for any finite-dimensional complex simple Lie algebra $\mathfrak{g}$, $H^2(\mathfrak{g}, U\mathfrak{g}) = 0$. Therefore, any formal deformation of $U\mathfrak{g}$ is trivial.

Let $B \subset A = U_h\mathfrak{g}$ be the preimage of $\mathfrak{g}$ under the map $A \to A/hA \simeq U\mathfrak{g}$. Then the short exact sequence of Lie algebras

$$0 \to U\mathfrak{g}\hat{\otimes}\mathfrak{m} \to B \to \mathfrak{g} \to 0,$$

where $\mathfrak{m} = h\mathbb{C}[[h]]$ is the maximal ideal in the local ring $\mathbb{C}[[h]]$, splits. Therefore, there is an isomorphism of Lie algebras $B \xrightarrow{\sim} \mathfrak{g} \oplus U\mathfrak{g}\hat{\otimes}\mathfrak{m}$. It follows that there is an isomorphism $U_h\mathfrak{g} \xrightarrow{\sim} U\mathfrak{g}[[h]]$ which is the identity modulo $\mathfrak{m}$.

It is clear that two such isomorphisms differ by an automorphism of $U\mathfrak{g}[[h]]$ which is the identity modulo $\mathfrak{m} = h\mathbb{C}[[h]]$. The well-known equality $H^1(\mathfrak{g}, U\mathfrak{g}) = 0$ implies that such automorphisms are inner. In particular, they act trivially on the center. Therefore, the center of $U_h\mathfrak{g}$ is canonically isomorphic to $Z[[h]]$, where Z is the center of $U\mathfrak{g}$.

Let $(U\mathfrak{g})^{\mathfrak{h}} = \{a \in U\mathfrak{g} \mid [\mathfrak{h}, a] = 0\}$ be the subspace of the invariants of the adjoint action of $\mathfrak{h}$ in $U\mathfrak{g}$.

EXERCISE 1.2.7. *Using the fact that $H^1\left(\mathfrak{h}, U\mathfrak{g}/(U\mathfrak{g})^{\mathfrak{h}}\right) = 0$, show that one can choose an isomorphism $\Phi : U_h\mathfrak{g} \to U\mathfrak{g}[[h]]$ so that $\Phi(\mathfrak{h}) \subset (U\mathfrak{g})^{\mathfrak{h}}[[h]]$.*

Consider the composition $\Psi = (\Phi \otimes \Phi) \circ \Delta \circ \Phi^{-1} : U\mathfrak{g}[[h]] \to (U\mathfrak{g} \otimes U\mathfrak{g})[[h]]$, where Δ is the comultiplication in $U_h\mathfrak{g}$. Since Ψ is an algebra homomorphism, $\Psi \bmod h$ is the usual comultiplication $\Delta_0 : U\mathfrak{g} \to U\mathfrak{g} \otimes U\mathfrak{g}$.

EXERCISE 1.2.8. *Show that there exists $F \in (U\mathfrak{g} \otimes U\mathfrak{g})[[h]]$ such that $F \equiv 1 \bmod h$ and $\Psi(a) = F\Delta_0(x)F^{-1}$ for any $a \in U\mathfrak{g}$.*

Hint: Use the fact that $H^1(\mathfrak{g}, U\mathfrak{g} \otimes U\mathfrak{g}) = 0$.

Suppose that $a \in \mathfrak{h}$. Since $\Delta(a) = a \otimes 1 + 1 \otimes a$, we have that $F\Delta_0^h(\Phi(a))F^{-1} = \Phi(a) \otimes 1 + 1 \otimes \Phi(a)$, where Δ_0^h is the comultiplication in $U\mathfrak{g}[[h]]$. Suppose that $\Phi(a) = a + u_1 h + u_2 h^2 + \cdots$, where $u_i \in U\mathfrak{g}$.

LEMMA 1.2.9. $u_i \in \mathfrak{h}$.

PROOF. We prove the lemma by induction. Indeed, we already know that $u_0 = a \in \mathfrak{h}$. Assume that $u_1, \ldots, u_{k-1} \in \mathfrak{h}$. Comparing the coefficients of h^k in

$$F\Delta_0^h(\Phi(a)) = (\Phi(a) \otimes 1 + 1 \otimes \Phi(a))F,$$

we get

$$\Delta_0(u_k) - u_k \otimes 1 - 1 \otimes u_k \in [\Delta_0(\mathfrak{h}), U\mathfrak{g} \otimes U\mathfrak{g}].$$

But since $\Phi(\mathfrak{h}) \in (U\mathfrak{g})^{\mathfrak{h}}[[h]]$, we have $u_k \in (U\mathfrak{g})^{\mathfrak{h}}$, hence

$$[\Delta_0(u_k) - u_k \otimes 1 - 1 \otimes u_k, \Delta_0(\mathfrak{h})] = 0.$$

Therefore, $\Delta_0(u_k) = u_k \otimes 1 + 1 \otimes u_k$, that is, $u_k \in \mathfrak{g}$. Thus, we see that $u_k \in \mathfrak{g} \cap (U\mathfrak{g})\mathfrak{h} = \mathfrak{h}$. □

By the lemma, $\Phi(\mathfrak{h}[[h]]) \subset \mathfrak{h}[[h]]$. Let $f : \mathfrak{h}^*[[h]] \to \mathfrak{h}^*[[h]]$ be the linear map conjugated to the restriction of Φ to $\mathfrak{h}[[h]]$. Since the spectra of the adjoint representations of $\mathfrak{h}$ in $U\mathfrak{g}$ and $U_h\mathfrak{g}$ are the same, f maps the root lattice into itself. Since $f \bmod h = \mathrm{id}$, this implies that $f = \mathrm{id}$, that is, $\Phi|_\mathfrak{h} = \mathrm{id}$. The theorem is proved. □

DEFINITION 1.2.10. Let A be a Hopf algebra with comultiplication Δ, antipode S, and counit ε. An element $g \in A$ is called *group-like* if $\Delta(g) = g \otimes g$.

EXERCISE 1.2.11. (1) Prove that if $g \in A$ is a group-like element, then $S(g) = g^{-1}$ and $\varepsilon(g) = 1$.
(2) Show that all group-like elements of a given Hopf algebra form a multiplicative group.

1.3. Existence and uniqueness of quantization. As soon as we have the concept of quantization of a given Lie bialgebra, there arises the natural question of when a quantization exists and when it is unique. We formulate here without proofs two results in this direction (see [EK96]).

THEOREM 1.3.1. *Any finite-dimensional Lie bialgebra $\mathfrak{g}$ over a field $\boldsymbol{k}$ of characteristic zero admits a quantization.*

THEOREM 1.3.2. *Let $\mathfrak{g}$ be a finite-dimensional simple complex Lie algebra. Fix a Cartan subalgebra $\mathfrak{h}$ in $\mathfrak{g}$ and a triangular decomposition $\mathfrak{g} = \mathfrak{n}_+ \oplus \mathfrak{h} \oplus \mathfrak{n}_-$ and consider the corresponding standard Lie bialgebra structure (cf. Example 2.1.8 of Chapter 1). Up to the substitution*
$$h \mapsto h + c_2 h^2 + c_3 h^3 + \cdots,$$
$U_h\mathfrak{g}$ *is the unique quantization A of $U\mathfrak{g}$ for which there exists a cocommutative Hopf subalgebra $C \subset A$ and an involutive algebra automorphism and coalgebra antiautomorphism $\gamma : A \to A$ such that:*
- *the natural mapping $C/hC \to U\mathfrak{g}$ is injective, with the image equal to $U\mathfrak{h}$;*
- $\gamma(C) = C$;
- $\gamma \bmod h$ *coincides with the Cartan involution $X_i^\pm \mapsto -X_i^\mp$, $i = 1, \ldots, n$, and $a \mapsto -a$ for $a \in \mathfrak{h}$.*

REMARK 1.3.3. In particular, we see that the quantization $U_h\mathfrak{g}$ of $\mathfrak{g}$ is essentially unique. We call $U_h\mathfrak{g}$ *the quantized universal enveloping algebra of $\mathfrak{g}$.*

2. QUE-algebras and R-matrices

In this section we introduce certain special types of Hopf algebras: almost-cocommutative, coboundary, quasi-triangular, and triangular ones. They are characterized by the existence of the so-called quantum R-matrix which is a distinguished invertible element of (a certain completion of) the second tensor power of the Hopf algebra considered. Some of their basic properties are investigated.

Then we study the quantum double construction, which is parallel on the quasi-classical level to the construction of the double Lie bialgebra given in Theorem 2.3.2 and Proposition 2.3.3 of Chapter 1. We observe in particular that it gives us an example of a quasi-triangular QUE-algebra. We deduce then that the quantized universal enveloping algebra $U_h\mathfrak{g}(A)$ introduced in the previous section is quasi-triangular as well and show that the quantum double construction can be used to obtain explicit formulas for the universal quantum R-matrix.

2.1. Types of Hopf algebras.

DEFINITION 2.1.1. Let A be a Hopf algebra over a field $\mathbf{k}$ of characteristic zero.

(1) An *almost-cocommutative* Hopf algebra over $\mathbf{k}$ is a pair (A, R), where $R \in A \otimes A$ is an invertible element such that

(2.1.1) $$R\Delta(a) = \Delta'(a)R$$

for any $a \in A$, where Δ is the comultiplication, $\Delta' = \sigma\Delta$ is the opposite comultiplication, and $\sigma(a \otimes b) = b \otimes a$.

(2) An almost-cocommutative Hopf algebra (A, R) is said to be *coboundary* if the following conditions are satisfied:

(2.1.2) $$R^{12}R^{21} = 1, \quad (\varepsilon \otimes \varepsilon)(R) = 1,$$
(2.1.3) $$R^{12}(\Delta \otimes \mathrm{id})(R) = R^{23}(\mathrm{id} \otimes \Delta)(R).$$

(3) An almost-cocommutative Hopf algebra (A, R) is said to be *quasi-triangular* if the following conditions are satisfied:

(2.1.4) $$(\Delta \otimes \mathrm{id})(R) = R^{13}R^{23},$$
(2.1.5) $$(\mathrm{id} \otimes \Delta)(R) = R^{13}R^{12}.$$

(4) A quasi-triangular Hopf algebra (A, R) is said to be *triangular* if $R^{12}R^{21} = 1$. Recall that for $R = \sum_k a_k \otimes b_k$, we denote $R^{12} = R \otimes 1$, $R^{23} = 1 \otimes R$, and $R^{13} = \sum_k a_k \otimes 1 \otimes b_k$.

DEFINITION 2.1.2. Let A be a Hopf algebra over a local commutative ring $\mathfrak{R}$ with the maximal ideal $\mathfrak{m}$ and the residue field $\mathbf{k} = \mathfrak{R}/\mathfrak{m}$ of characteristic zero. Suppose that A is complete in the $\mathfrak{m}$-adic topology.

(1) An *almost-cocommutative* Hopf algebra is a pair (A, R), where $R \in A\hat{\otimes}A$ is an invertible element which satisfies (2.1.1) and such that $R = 1 \bmod \mathfrak{m}$.

(2) An almost-cocommutative Hopf algebra (A, R) is said to be *coboundary* if R satisfies (2.1.2)–(2.1.3).

(3) An almost-cocommutative Hopf algebra (A, R) is said to be *quasi-triangular* if R satisfies (2.1.4), (2.1.5).

(3) A quasi-triangular Hopf algebra (A, R) is said to be *triangular* if $R^{12}R^{21} = 1$.

DEFINITION 2.1.3. In all the cases considered in Definitions 2.1.1 and 2.1.2 R is called the *quantum R-matrix* for A.

REMARK 2.1.4. Note that given a quasi-triangular (coboundary, triangular) Hopf algebra, the quantum R-matrix is not unique. For example, is R satisfies the respective conditions, $\tilde{R} = \left(R^{21}\right)^{-1}$ also satisfies the corresponding conditions.

Now we are going to discuss some basic properties of the quantum R-matrix which follow directly from the above definitions. Throughout the rest of the subsection we give proofs only in the case of Hopf algebras over $\mathbf{k}$. The corresponding results for Hopf algebras over a complete local commutative ring $\mathfrak{R}$ with the residue field $\mathbf{k}$ can be deduced from the following observation.

EXERCISE 2.1.5. Suppose that (A, R) is an almost-cocommutative (resp. coboundary, quasi-triangular, triangular) Hopf algebra over a local commutative ring $\mathfrak{R}$ with the maximal ideal $\mathfrak{m}$ and the residue field $\mathbf{k} = \mathfrak{R}/\mathfrak{m}$. Suppose also that A is complete in the $\mathfrak{m}$-adic topology.

Show that $(A/\mathfrak{m}^n A, R \bmod \mathfrak{m}^n)$ is an almost-cocommutative (resp. coboundary, quasi-triangular, triangular) Hopf algebra over $\mathbf{k}$ for any $n = 1, 2, \ldots$.

PROPOSITION 2.1.6. *Given an almost-cocommutative Hopf algebra (A, R) and the dual Hopf algebra A^* (see Definition 2.2.7 of Chapter 1), consider the map $\psi : A^* \to A$ given by $\psi : f \mapsto (f \otimes \mathrm{id})(R)$, where $f \in A^*$. Then (A, R) is quasi-triangular if and only if $\psi : A^* \to A$ is an algebra homomorphism and a coalgebra antihomomorphism.*

PROOF. Suppose that (A, R) is a quasi-triangular Hopf algebra. Then

$$\begin{aligned}(2.1.6) \quad \psi(fg) &= (fg \otimes \mathrm{id})(R) = (f \otimes g \otimes \mathrm{id})(\Delta \otimes \mathrm{id})(R) \\ &= (f \otimes g \otimes \mathrm{id})\left(R^{13}R^{23}\right) = \psi(f)\psi(g), \\ (2.1.7) \quad \Delta(\psi(f)) &= (f \otimes \mathrm{id} \otimes \mathrm{id})(\mathrm{id} \otimes \Delta(R)) = (f \otimes \mathrm{id} \otimes \mathrm{id})\left(R^{13}R^{12}\right) \\ &= (\Delta'(f) \otimes \mathrm{id} \otimes \mathrm{id})\left(R^{13}R^{24}\right) = \psi(\Delta'(f)).\end{aligned}$$

On the other hand, we see that if ψ is an algebra homomorphism and a coalgebra antihomomorphism, then (2.1.6) and (2.1.7) are satisfied for any $f, g \in A$. It follows that both (2.1.4) and (2.1.5) are satisfied, so that (A, R) is a quasi-triangular Hopf algebra. □

PROPOSITION 2.1.7. *Suppose that (A, R) is a quasi-triangular Hopf algebra. Then the following equation holds:*

(2.1.8) $\quad R^{12}R^{13}R^{23} = R^{23}R^{13}R^{12}(S \otimes \mathrm{id})(R) = R^{-1} = (\mathrm{id} \otimes S^{-1})(R),$
(2.1.9) $\quad\quad\quad\quad\quad\quad\quad\quad\quad (S \otimes S)(R) = R,$
(2.1.10) $\quad\quad\quad\quad\quad\quad\quad\quad (\varepsilon \otimes \mathrm{id})(R) = 1 = (\mathrm{id} \otimes \varepsilon)(R).$

PROOF. By (2.1.1), (2.1.4), (2.1.5), we have $R^{12}R^{13}R^{23} = R^{12}(\Delta \otimes \mathrm{id})(R) = (\Delta' \otimes \mathrm{id})(R)R^{12} = R^{23}R^{13}R^{12}$. Applying $\varepsilon \otimes \mathrm{id} \otimes \mathrm{id}$ to (2.1.4), we get $(\varepsilon \otimes \mathrm{id})(R) = 1$. Applying $\mathrm{id} \otimes \mathrm{id} \otimes \varepsilon$ to (2.1.5), we get $(\mathrm{id} \otimes \varepsilon)(R) = 1$. Let $m : A \otimes A \to A$ be the multiplication map $m : x \otimes y \mapsto xy$. Then

$$\begin{aligned}R \cdot (S \otimes \mathrm{id})(R) &= (m \otimes \mathrm{id})(\mathrm{id} \otimes S \otimes \mathrm{id})\left(R^{13}R^{23}\right) \\ &= (m \otimes \mathrm{id})(\mathrm{id} \otimes S \otimes \mathrm{id})(\Delta \otimes \mathrm{id})(R) \\ &= (\varepsilon \otimes \mathrm{id})(R) = 1,\end{aligned}$$

that is, $(S \otimes \mathrm{id})(R) = R^{-1}$.

Consider the Hopf algebra A° which is the algebra A with the opposite comultiplication and the skew antipode S^{-1}. Then it is also quasi-triangular, with the quantum R-matrix R^{21}. Therefore, $(S^{-1} \otimes \mathrm{id})\left(R^{21}\right) = \left(R^{21}\right)^{-1}$, that is $(\mathrm{id} \otimes S^{-1})(R) = R^{-1}$. Since $(\mathrm{id} \otimes S^{-1})(R) = (S \otimes \mathrm{id})(R)$, it follows that $(S \otimes S)(R) = R$. □

DEFINITION 2.1.8. The equation (2.1.8) is called the *quantum Yang–Baxter equation*.

EXERCISE 2.1.9. Prove the following statements.
(1) Any triangular Hopf algebra is a coboundary Hopf algebra.
(2) Suppose that (A, R) is a coboundary Hopf algebra and the quantum Yang–Baxter equation (2.1.8) holds. Then (A, R) is triangular.
(3) Given a quasi-triangular Hopf algebra (A, R), consider the element $\tilde{R} = \left(R^{12}R^{21}\right)^{-1/2} R^{12}$. Then $(A, \tilde{R})$ is a coboundary Hopf algebra.

DEFINITION 2.1.10. A Hopf algebra A with comultiplication Δ is called *cocommutative* if $\Delta = \Delta'$, where $\Delta' = \sigma \Delta$ is the opposite comultiplication (σ is the permutation of the tensor factors in $A \otimes A$).

EXERCISE 2.1.11. Let (A, Δ_0) be a cocommutative Hopf algebra with comultiplication Δ_0. Suppose that there is an invertible element $\Phi \in A \otimes A$ such that

$$\Phi^{12}(\Delta_0 \otimes \mathrm{id})(\Phi) = \Phi^{23}(\mathrm{id} \otimes \Delta_0)(\Phi).$$

Consider a new comultiplication Δ on A and an element $R \in A \otimes A$ given by

(2.1.11) $\qquad \Delta(a) = \Phi \Delta_0(a) \Phi^{-1}, \qquad R = \Phi^{21}\left(\Phi^{12}\right)^{-1}.$

Show that (A, Δ, R) is a triangular Hopf algebra.

2.2. Double Hopf algebras. Important examples of quasi-triangular Hopf algebras are provided by the (quantum) double construction. We will see also that in the case of a QUE-algebra the quasi-classical analog of the double QUE-algebra is the double Lie bialgebra defined in Definition 2.3.4 of Chapter 1.

Recall that, given a finite-dimensional vector space V, the canonical element of $V \otimes V^*$ is by definition the element $t = \sum_k e_k \otimes e^k \in V \otimes V^*$, where $\{e_k\}$ and $\{e^k\}$ are dual bases of V and V^*. Note that t does not depend on the choice of the bases.

THEOREM 2.2.1. *Let A be a finite-dimensional Hopf algebra over a field $\mathbf{k}$ of characteristic zero (then the dual space A^* carries the dual Hopf algebra structure given by (2.2.6)–(2.2.8)). Suppose that A° is the Hopf algebra isomorphic to A^* as algebra, whose comultiplication is the opposite comultiplication in A^*.*

Then there exists a unique quasi-triangular Hopf algebra $(\mathcal{D}(A), R)$ which satisfies the following conditions:
(1) *there exist embeddings $i_1 : A \hookrightarrow \mathcal{D}(A)$ and $i_2 : A^\circ \hookrightarrow \mathcal{D}(A)$ of Hopf algebras;*
(2) *the linear map $A \otimes A^\circ \to \mathcal{D}(A)$ given by the multiplication in $\mathcal{D}(A)$ is bijective;*
(3) *R is the image of the canonical element of $A \otimes A^\circ$ by the embedding $i_1 \otimes i_2 : A \otimes A^\circ \to \mathcal{D}(A) \otimes \mathcal{D}(A)$.*

DEFINITION 2.2.2. The quasi-triangular Hopf algebra $(\mathcal{D}(A), R)$ described in Theorem 2.2.1 is called the *double Hopf algebra* of A.

PROOF OF THEOREM 2.2.1. The following explicit construction proves the existence part of the theorem.

THE QUANTUM DOUBLE CONSTRUCTION. As a vector space $\mathcal{D}(A)$ can be identified with $A \otimes A^\circ$ by $a \otimes b \mapsto ab$, and its Hopf algebra structure will be completely determined as soon as one knows the multiplication map $A^\circ \otimes A \to \mathcal{D}(A)$. An intrinsic formula for it looks as follows:

$$
(2.2.1) \qquad A^\circ \otimes A \xrightarrow{\bar\Delta} (A^\circ \otimes A)^{\otimes 2} \xrightarrow{S \otimes \mathrm{id}^{\otimes 3}} (A^\circ \otimes A)^{\otimes 2}
$$

$$
\begin{array}{c}
\downarrow \\
\mathcal{D}(A) \longleftarrow A \otimes A^\circ \xleftarrow{\mathrm{id}^{\otimes 2} \otimes \langle\,,\,\rangle} (A \otimes A^\circ)^{\otimes 2}
\end{array}
$$

with vertical maps $\sigma(\langle\,,\,\rangle \otimes \mathrm{id}^{\otimes 2})$ into $A^\circ \otimes A$ and $\bar\Delta$.

where $\bar\Delta = \sigma_{23}(\Delta \otimes \Delta)$, σ stands for the permutation of the tensor factors, and $\langle\,,\,\rangle$ is the natural pairing between A and A^*. (The maps without any indication in the diagram (2.2.1) are multiplication maps.)

EXERCISE 2.2.3. Check that this defines a Hopf algebra structure on $\mathcal{D}(A)$.

There is another, "coordinate," way of describing the same structure on $\mathcal{D}(A)$ (the reader can easily see their equivalence). Namely, let $\{e_k\}$ and $\{e^k\}$ be dual bases of A and A°, respectively.

EXERCISE 2.2.4. Prove that

$$
(2.2.2) \qquad e_r e^t = \sum_{l,n,p} \mu_r^{ljn} m^t_{pkl} s^p_n e^k e_j,
$$

$$
(2.2.3) \qquad e^t e_r = \sum_{l,n,p} \mu_r^{njl} m^t_{lkp} s^p_n e_j e^k.
$$

Here μ, m, and s are the matrices of the comultiplication map $A \to A \otimes A \otimes A$, the multiplication map $A \otimes A \otimes A \to A$, and the "skew" antipode $S^{-1}: A \to A$ respectively.

The conditions (1) and (2) of the theorem imply that $\mathcal{D}(A) \simeq A \otimes A^\circ$ as a vector space and that the coalgebra structure is defined uniquely by the multiplication in $\mathcal{D}(A)$ and the coalgebra structures on A and A°. The third condition is responsible for the fact that the multiplication is unique and given by the formulas (2.2.2) and (2.2.3). This proves the uniqueness part of the theorem. $\square$

2.3. The quantum double and the universal quantum R-matrix. Let A be a QUE-algebra. Then the construction of the quantum double should be modified, since the dual Hopf algebra A^*, consisting of $\mathbf{k}[[h]]$-linear maps $A \to \mathbf{k}[[h]]$ which are continuous in the h-adic topology, is not a QUE-algebra.

DEFINITION 2.3.1. A *quantum formal series Hopf algebra* (QFSH-algebra for short) is a topological Hopf algebra B over $\mathbf{k}[[h]]$ such that

(1) B/hB is isomorphic to the topological algebra $\mathbf{k}[[u_1, u_2, \dots]]$.
(2) As a topological $\mathbf{k}[[h]]$-module, B is isomorphic to $\mathbf{k}[[h]]^I$ for some set I.

REMARK 2.3.2. Informally speaking, a QFSH-algebra B is a quotient algebra of an algebra of noncommutative formal power series (over $\mathbf{k}[[h]]$) in $u_1, u_2, \dots$ over the ideal generated by the elements of the form $[u_i, u_j] - \psi_{ij}(u_1, u_2, \dots; h)$ where $\psi(0, 0, \dots; h) = 0$. There are also "compatibility relations" between ψ_{ij} (since B is free over $\mathbf{k}[[h]]$). The comultiplication $\Delta(u_i)$ can be expressed in terms of a series $F_i(u_1 \otimes 1, u_2 \otimes 1, \dots; 1 \otimes u_1, 1 \otimes u_2, \dots)$.

EXERCISE 2.3.3. Given a QUE-algebra A, prove that the dual Hopf algebra A^* is a QFSH-algebra.

PROPOSITION 2.3.4. *Let A be a quantization of a Lie bialgebra $(\mathfrak{g}, \varphi)$. Consider the Hopf subalgebra $A^\sharp$ in A^* which is the h-adic completion of $\bigoplus_{n=0}^{\infty} h^{-n} \mathfrak{m}^n$, where $\mathfrak{m} \subset A^*$ is the ideal topologically generated by the elements $x_i \in A$ $(i = 1, 2, \ldots)$ such that $x_i = u_i \bmod h$. Then $A^\sharp$ is a quantization of the dual Lie bialgebra $\mathfrak{g}^*$.*

PROOF. We give a sketch of the proof leaving the details as an exercise to the reader.

Step 1. It follows from the definition that $A^\sharp$ is a topologically free $\mathbf{k}[[h]]$-module topologically generated by $t_i = h^{-1} x_i$ $(i = 1, 2, \ldots)$.

Step 2. (Exercise) Prove that the complete set of relations between the elements t_i is given by $[t_i, t_j] = h^{-1} \psi_{ij}(ht_1, ht_2, \ldots)$, where ψ_{ij} means the same as in Remark 2.3.2.

Step 3. It follows from the previous step that $A^\sharp / h A^\sharp$ is a co-Poisson Hopf algebra which is canonically isomorphic to the co-Poisson universal enveloping algebra of the Lie bialgebra dual to $(\mathfrak{g}, \varphi)$. This completes the proof. □

Note that the canonical element is well defined as an element of $A \otimes A^\sharp$.

DEFINITION 2.3.5. The *quantum double* $\mathcal{D}(A)$ of a QUE-algebra A is defined in the same way as in Theorem 2.2.1 and Definition 2.2.2, where A° is understood as the Hopf algebra opposite to $A^\sharp$.

EXERCISE 2.3.6. (1) Check that the quantum double $\mathcal{D}(A)$ of a QUE-algebra A exists and can be obtained using the quantum double construction described in Section 2.2.

(2) Prove that if A is a quantization of a Lie bialgebra $\mathfrak{g}$, then $\mathcal{D}(A)$ is a quantization of the double Lie bialgebra $\mathcal{D}(\mathfrak{g})$ (cf. Proposition 2.3.3 of Chapter 1).

Let $\mathfrak{g}$ be a complex simple Lie algebra with the standard Lie bialgebra structure, and $U_h \mathfrak{g}$ its quantization. Define $U_h \mathfrak{b}_\pm$ as in Remark 1.2.5.

PROPOSITION 2.3.7. *There is an isomorphism of Hopf algebras $f : (U_h \mathfrak{b}_+)^\circ \xrightarrow{\sim} U_h \mathfrak{b}_-$ such that the map $\bar{f} = f \bmod h$ defines an isomorphism of Lie algebras $\bar{f} : \mathfrak{b}_+^* \xrightarrow{\sim} \mathfrak{b}_-$ and an antiisomorphism of Lie bialgebras $\bar{f}^* : \mathfrak{b}_-^* \to \mathfrak{b}_+$.*

PROOF. We start with an obvious remark. The elements of the form $x_{\bar{n}, \bar{l}} = E_{i_1}^{n_1} \cdots E_{i_p}^{n_p} H_1^{l_1} \cdots H_n^{l_n}$ (E_i as in (1.2.9)) generate the topological $\mathbb{C}[[h]]$-module $U_h \mathfrak{b}_+$ for nonnegative n_i and l_i. Therefore, $\varphi = \psi$ in $(U_h \mathfrak{b}_+)^*$ if and only if they coincide on all those elements.

Define ξ_i, η_i $(i = 1, 2, \ldots, n)$ as the following linear functional $U_h \mathfrak{b}_+ \to \mathbb{C}[[h]]$:

(2.3.1) $\qquad \xi_i(H_i) = 1 \quad \text{and} \quad \xi_i(x_{\bar{n}, \bar{l}}) = 0$ otherwise,

(2.3.2) $\qquad \eta_i(E_i) = 1 \quad \text{and} \quad \eta_i(x_{\bar{n}, \bar{l}}) = 0$ otherwise.

2. QUE-ALGEBRAS AND R-MATRICES

LEMMA 2.3.8. *The following relations hold in* $(U_h\mathfrak{b}_+)^*$:

(2.3.3) $$[\xi_i, \xi_j] = 0,$$

(2.3.4) $$[\xi_i, \eta_j] = -\delta_{ij}\frac{h}{2}\eta_j,$$

(2.3.5) $\Delta(\xi_i) = \xi_i \otimes 1 + 1 \otimes \xi_i, \qquad \Delta(\eta_i) = \eta_i \otimes 1 + e^{\sum_{k=1}^n (\alpha_i, \alpha_k)\xi_k} \otimes \eta_i,$

(2.3.6) $$(\mathrm{ad}_{\eta_i})^{1-a_{ij}}(\eta_j) = 0, \quad i \neq j,$$

where ad_{η_i} stands for the quantum adjoint action of η_i.

PROOF. The proofs of (2.3.3)–(2.3.5) are similar, so that we will give the proof of (2.3.4) only. We want to compare $\langle \xi_i \eta_j, x_{\bar{n},\bar{l}} \rangle$ and $\langle \eta_j \xi_i, x_{\bar{n},\bar{l}} \rangle$. Note that

$$\begin{aligned}\langle \xi_i \eta_j, x_{\bar{n},\bar{l}} \rangle &= \langle \xi_i \otimes \eta_j, \Delta(x_{\bar{n},\bar{l}}) \rangle \\ &= \langle \xi_i \otimes \eta_j, \Delta(E_{i_1})^{n_1} \cdots \Delta(E_{i_p})^{n_p} \Delta(H_1)^{l_1} \cdots \Delta(H_n)^{l_n} \rangle.\end{aligned}$$

We have that

(2.3.7) $$\Delta(E_i)^m = \sum_{s=0}^m \binom{m}{s}_{q_i^{-2}} E_i^s q^{-(m-s)H_i} \otimes E_i^{m-s},$$

where $\binom{m}{s}_t = \frac{(m)_t!}{(s)_t!(m-s)_t!}$ and $(n)_t! = (1)_t \cdots (n)_t$, $(n)_t = \frac{1-t^n}{1-t}$. To check (2.3.7), note that

$$(E_i \otimes 1)\left(q^{-H_i} \otimes E_i\right) = q^{(\alpha_i, \alpha_i)}\left(q^{-H_i} \otimes E_i\right)(E_i \otimes 1).$$

Then use (1.2.10) and the q-binomial formula

$$(a+b)^m = \sum_{s=0}^m \binom{m}{s}_{q^{-1}} a^s b^{m-s}$$

which holds once $ab = qba$.

Formulas (2.3.1) and (2.3.2) show that we are interested only in those terms in $\Delta(E_{i_1})^{n_1} \cdots \Delta(H_n)^{l_n}$ which are of the form $c_{ij} H_i \otimes E_j$, where $c_{ij} \in \mathbb{C}[[h]]$. Since all the relations in $U_h\mathfrak{b}_+$ are homogeneous, (2.3.7) implies that $\langle \xi_i \eta_j, x_{\bar{n},\bar{l}} \rangle = 0$ unless either one of the following conditions is satisfied:

(1) $l_i = 0$ for any i, there exists a unique $1 \leq t \leq p$ such that $i_t = j$, and for that t we have that $n_t = 1$;
(2) there exists a unique $1 \leq t \leq p$ such that $i_t = j$, we have for that t that $n_t = 1$ and $l_{i_t} = 1$, and $l_i = 0$ if $i \neq i_t$.

In the first case we have that

(2.3.8) $$\langle \xi_i \otimes \eta_j, E_j \otimes 1 + q^{-H_j} \otimes E_j \rangle = -\delta_{ij}\frac{h}{2}.$$

In the second case we have that

(2.3.9) $$\langle \xi_i \otimes \eta_j, \left(E_j \otimes 1 + q^{-H_j} \otimes E_j\right)(H_i \otimes 1 + 1 \otimes H_i) \rangle = 1.$$

Similarly, we can compute $\langle \eta_j \xi_i, x_{\bar{n},\bar{l}} \rangle$. We still have only two nontrivial cases:

(2.3.10) $$\langle \eta_j \otimes \xi_i, E_j \otimes 1 + q^{-H_j} \otimes E_j \rangle$$

and

(2.3.11) $$\langle \eta_j \otimes \xi_i, \left(E_j \otimes 1 + q^{-H_j} \otimes E_j\right)(H_i \otimes 1 + 1 \otimes H_i) \rangle.$$

But now we see that (2.3.10) is always equal to zero, while (2.3.11) is always equal to 1. Therefore, we get that

$$\langle \xi_i \eta_j - \eta_j \xi_i, x_{\bar{n},\bar{l}} \rangle = -\delta_{ij} \frac{h}{2} \langle \eta_j, x_{\bar{n},\bar{l}} \rangle,$$

which proves (2.3.4). As we have mentioned, the proofs of (2.3.3) and (2.3.5) are similar.

To prove (2.3.6), introduce the generators $\bar{\xi}_i = \sum_{k=1}^{n} (\alpha_i, \alpha_k) \xi_k$. By (2.3.3)–(2.3.5), we have that

(2.3.12) $\qquad\qquad [\bar{\xi}_i, \bar{\xi}_j] = 0,$

(2.3.13) $\qquad\qquad [\bar{\xi}_i, \eta_j] = -\delta_{ij}(\alpha_j, \alpha_j)\eta_j,$

(2.3.14) $\quad \delta(\bar{\xi}_i) = \bar{\xi}_i \otimes 1 + 1 \otimes \bar{\xi}_i, \qquad \Delta(\eta_i) = \eta_i \otimes 1 + e^{\bar{\xi}_i} \otimes \eta_i.$

Consider the Hopf subalgebra A_i° of $(U_h \mathfrak{b}_+)^\circ$ generated by $h^{-1}\bar{\xi}_i$ and $h^{-1}\eta_i$. It is isomorphic to the Hopf subalgebra $U_h \mathfrak{b}_-^i$ in $U_h \mathfrak{b}_+$ generated by H_i and F_i. Fix i and j such that $i \neq j$. Consider the following topologically free $\mathbb{C}[[h]]$-submodule of $(U_h \mathfrak{b}_+)^*$:

(2.3.15) $\qquad\qquad V_{ij} = \text{Span}\{\eta_j, \text{ad}_{\eta_i}(\eta_j), \ldots, (\text{ad}_{\eta_i})^{-a_{ij}}(\eta_j)\}.$

Then (2.3.12)–(2.3.14) show that A_i° acts on V_{ij} and, therefore, V_{ij} is a $U_h \mathfrak{b}_-^i$-module of finite rank. It is easy to see that any such module admits a natural structure of a $U_h \mathfrak{sl}(2)_i$-module (see Remark 1.2.5 for the notation). The rank of V_{ij} as a free $\mathbb{C}[[h]]$-module is equal to the complex dimension of the corresponding $\mathfrak{sl}(2)_i$-module. Therefore, $(\text{ad}_{\eta_i})^{1-a_{ij}} \eta_j \in V_{ij}$. Since V_{ij} is graded with respect to the commutative Lie algebra generated by $\bar{\xi}_1, \ldots, \bar{\xi}_n$ and the degrees of $\eta, \text{ad}_{\eta_i}(\eta_j), \ldots, (\text{ad}_{\eta_i})^{1-a_{ij}}(\eta_j)$ are all different, we conclude that $(\text{ad}_{\eta_i})^{1-a_{ij}}(\eta_j) = 0$. The lemma is proved. $\square$

Now let $\bar{c} = (c_1, \ldots, c_n)$ be a sequence of elements from $\mathbb{C}[[h]]$ such that $c_i(0) \neq 0$ for $1 \leq i \leq r$. Define a homomorphism of $\mathbb{C}[[h]]$-modules $\Psi_{\bar{c}} : U_h \mathfrak{b}_- \to (U_h \mathfrak{b}_+)^*$ such that

(2.3.16) $\qquad\qquad \Psi_{\bar{c}}(F_j) = c_j^{-1} h^{-1} \eta_j,$

(2.3.17) $\qquad\qquad \Psi_{\bar{c}}(H_j) = 2h^{-1}\bar{\xi}_j.$

Lemma 2.3.8 together with (2.3.12)–(2.3.14) shows that $\Psi_{\bar{c}} : U_h \mathfrak{b}_- \to (U_h \mathfrak{b}_+)^\circ$ is a homomorphism of Hopf algebras. We claim that this is an isomorphism.

Since both $U_h \mathfrak{b}_-$ and $(U_h \mathfrak{b}_+)^\circ$ are topologically free over $\mathbb{C}[[h]]$, it suffices to verify whether $\Psi_{\bar{c}}$ is an isomorphism modulo h. But in the case $h = 0$ it is equal to the composition of the isomorphism of co-Poisson Hopf algebras arising from the duality $\mathfrak{b}_+^* \simeq \mathfrak{b}_-$ of Lie bialgebras and the automorphism $U\mathfrak{b}_- \xrightarrow{\sim} U_h \mathfrak{b}_-$ given by the dilation $X_i^- \mapsto c_i^{-1}(0) X_i^-$, $H_i \mapsto H_i$ ($i = 1, 2, \ldots, n$). Therefore, $\Psi_{\bar{c}}$ is indeed an isomorphism.

In order to get the map f as stated in the proposition, take $\bar{c} = (c_1, \ldots, c_n)$ such that $c_i(0) = 1$ for any $i = 1, 2, \ldots, n$ and put $f = \Psi_{\bar{c}}^{-1}$. The proposition is proved. $\square$

COROLLARY 2.3.9. *There is an epimorphism* $\Phi : \mathcal{D}(U_h \mathfrak{b}_+) \to U_h \mathfrak{g}$ *of Hopf algebras such that* $\Phi(a \otimes 1) = a$ *for any* $a \in U_h \mathfrak{b}_+$ *and* $\Phi(1 \otimes b) = \Psi_{\bar{c}}(b)$ *for* $b \in (U_h \mathfrak{b}_+)^\circ$, *where* $\bar{c} = (c_1, \ldots, c_n)$ *and* $c_i = 1 - q_i^{-2}$.

PROOF. After the proof of Proposition 2.3.7, the only thing left to check is the compatibility of Φ with the commutation relations between ξ_i, η_i and E_i, H_i ($i = 1, 2, \ldots, n$). But this is an immediate consequence of (2.2.1). In particular, the proposition implies that $c_i = 1 - q_i^{-2}$. □

COROLLARY 2.3.10. *The image $R = \Phi(\bar{R})$ of the canonical element $\bar{R} \in \mathcal{D}(U_h \mathfrak{b}_+)$ has the following properties:*

(2.3.18) $$R \in U_h\mathfrak{b}_+ \hat{\otimes} U_h\mathfrak{b}_- \text{ is invertible,}$$

(2.3.19) $$R\Delta(a)R^{-1} = \Delta'(a), \ a \in U_h\mathfrak{g},$$

(2.3.20) $$(\Delta \otimes \mathrm{id})(R) = R^{13}R^{23}, \ (\mathrm{id} \otimes \Delta)(R) = R^{13}R^{12}.$$

In particular, $(U_h\mathfrak{g}, R)$ is a quasi-triangular Hopf algebra.

DEFINITION 2.3.11. The quantum R-matrix defined in Corollary 2.3.10 is called the *universal R-matrix* for $U_h\mathfrak{g}$.

PROPOSITION 2.3.12. *Let $\mathfrak{g} = \mathfrak{sl}(2, \mathbb{C})$. Then the universal quantum R-matrix R is given by*

(2.3.21) $$R = \exp_{q^{-2}}\left((1 - q^{-2})E \otimes F\right) q^{\frac{1}{2}H \otimes H},$$

where $\exp_t(x) = \sum_{k=0}^{\infty} \frac{x^k}{(k)_t!}$ is the so-called q-exponent function. Here $(k)_t! = (1)_t(2)_t \cdots (k)_t$, where $(k)_t = \frac{1-t^k}{1-t}$.

PROOF. We leave it to the reader to verify that the canonical element $\bar{R}$ in $\mathcal{D}(U_h\mathfrak{b}_+)$ is equal to

$$\sum_{m,n=0}^{\infty} \left(E^m \otimes \frac{\eta^m}{(m)_{q^{-2}}!}\right)\left(H^n \otimes \frac{\xi^n}{n!}\right)$$

in the notation of (2.3.1), (2.3.2) (here $i = 1$). Then (2.3.21) follows from Corollaries 2.3.9 and 2.3.10. □

We will perform the computations for the general case in Chapter 4. But for now we describe the general form of the universal quantum R-matrix in the following proposition.

PROPOSITION 2.3.13. *Let $R \in U_h\mathfrak{g}\hat{\otimes} U_h\mathfrak{g}$ be a quantum R-matrix defined in Corollary 2.3.10. Then*

(2.3.22) $$R = \sum_{\beta \in \mathbb{Z}_+^n} \exp h\left(\frac{t_0}{2} + \frac{1}{4}(H_\beta \otimes 1 - 1 \otimes H_\beta)\right) P_\beta,$$

where:

(1) $t_0 \in \mathfrak{h} \otimes \mathfrak{h}$ is the canonical element corresponding to the invariant bilinear form on $\mathfrak{h}$;

(2) for $\beta = (\beta_1, \ldots, \beta_n)$, we have $H_\beta = \sum_{i=1}^n \beta_i H_i$;

(3) P_β is a noncommutative polynomial in $u_i = X_i^+ \otimes 1$ and $v_i = 1 \otimes X_i^-$, $i = 1, 2, \ldots, n$, with coefficients in $\mathbb{C}[[h]]$, such that $\deg_{u_i} P_\beta = \deg_{v_i} P_\beta = \beta_i$;

(4) $P_\beta \equiv 0 \mod h^{k_\beta}$ and $P_\beta \not\equiv 0 \mod h^{k_\beta - 1}$, where k_β is the minimal number n such that H_β can be expressed as a sum of n positive roots.

The properties (2.1.1), (2.1.4), (2.1.5), and (2.3.15) define R uniquely.

PROOF. It follows from (2.1.1) that
$$[R, a \otimes 1 + 1 \otimes a] = 0 \tag{2.3.23}$$
for any $a \in \mathfrak{h}$. Therefore, we have that
$$R = \sum_{\beta \in \mathbb{Z}_+^n} R_\beta, \tag{2.3.24}$$
where $R_\beta \in U_h\mathfrak{b}_+ \hat{\otimes} U_h\mathfrak{b}_-$ and $[a \otimes 1, R_\beta] = (H_\beta, a)$, $[1 \otimes a, R_\beta] = -(H_\beta, a)$, where $(\,,\,)$ is the invariant bilinear form on $\mathfrak{h}$. Then (2.1.4) and (2.1.5) imply that $P_0 = 1$. Consider (2.1.1) with $a = X_i^\pm$. Then for $\beta = (0, \ldots, 1, \ldots, 0)$ (1 at the ith place), we obtain
$$P_\beta = h \exp\left(-\frac{h}{4}(H_i, H_i)\right) X_i^+ \otimes X_i^-. \tag{2.3.25}$$

Using (2.1.4), we can find all the polynomials P_β by induction. Then for P_β, we have an equation of the form
$$(\Delta \otimes \mathrm{id})(P_\beta) \;-\; \exp\left(\frac{h}{4}(1 \otimes H_\beta \otimes 1)\right) P_\beta^{13} \tag{2.3.26}$$
$$-\; \exp\left(-\frac{h}{4}(H_\beta \otimes 1 \otimes 1)\right) P_\beta^{23} = x_\beta,$$
where x_β is known and P_β^{13}, P_β^{23} are defined similarly to R^{13} and R^{23}.

It is easy to see that there exists a solution to (2.3.26) and it satisfies the properties (3) and (4) of the proposition. The uniqueness of P_β follows from the lemma. Introduce the $\mathbb{Z}_+^n$-grading in $U_h\mathfrak{g}$ such that $\deg a = 0$ for $a \in \mathfrak{h}$ and $\deg X_i^\pm = (0, \ldots, \pm 1, \ldots, 0)$ (± 1 is at the ith place).

LEMMA 2.3.14. *Let $x \in U_h\mathfrak{g}$ be a homogeneous element of degree $\beta = (\beta_1, \beta_2, \ldots, \beta_n)$ such that $\sum_{i=1}^n \beta_i > 1$ and $\beta_i \geq 0$ ($i = 1, 2, \ldots, n$). If*
$$\Delta(x) = x \otimes \exp\left(\frac{hH_\beta}{4}\right) + \exp\left(-\frac{hH_\beta}{4}\right) \otimes x,$$
then $x = 0$.

PROOF. Let $\bar{x}$ be the image of x in $U_h\mathfrak{g}/hU_h\mathfrak{g} \simeq U\mathfrak{g}$. It suffices to show that $\bar{x} = 0$, since we can always replace x by $h^{-n}x$ such that n is the minimal integer for which $x \in h^n U_h\mathfrak{g}$ but $x \notin h^{n+1} U_h\mathfrak{g}$. We have that $\Delta_0(\bar{x}) = \bar{x} \otimes 1 + 1 \otimes \bar{x}$, where Δ_0 is the comultiplication in $U\mathfrak{g}$. Hence $\bar{x} \in \mathfrak{g}$. Our assumptions imply that $\bar{x} \in \mathfrak{b}_+$.

Let $\varphi : \mathfrak{g} \to \mathfrak{g} \otimes \mathfrak{g}$ be the cocommutator. Then
$$\varphi(\bar{x}) = h^{-1}(\Delta(x) - \Delta'(x)) \bmod h = \frac{1}{2}(\bar{x} \wedge H_\beta).$$

Let $y, z \in \mathfrak{b}_+^* \simeq \mathfrak{b}_-$ be homogeneous elements of nonzero degree. Then $(\bar{x}, [y, z]) = (\varphi(\bar{x}), y \otimes z)$. Here we denote by $(\,,\,)$ the invariant bilinear form on $\mathfrak{g}$, corresponding to the given invariant bilinear form on $\mathfrak{h}$. But since $\beta_1 + \cdots + \beta_n > 1$, this is possible if and only if $\bar{x} = 0$. The lemma is proved. □

Hence, Proposition 2.3.13 is also proved. □

EXERCISE 2.3.15. (1) Show that there exists a unique $\mathbb{C}$-linear algebra automorphism σ of $U_h\mathfrak{g}$ such that
$$\sigma(h) = -h, \quad \sigma\left(X_i^\pm\right) = X_i^\pm, \quad \sigma(H_i) = H_i. \tag{2.3.27}$$

Hint: use the fact that only even powers of h occur in the defining relations (1.2.1)–(1.2.3) among the generators of $U_h\mathfrak{g}$.

(2) Show that there exists a unique $\mathbb{C}[[h]]$-linear algebra antiautomorphism θ of $U_h\mathfrak{g}$ such that

(2.3.28) $$\theta\left(X_i^{\pm}\right) = X_i^{\mp}, \quad \theta(H_i) = H_i.$$

PROPOSITION 2.3.16. *Let R be the universal quantum R-matrix of $U_h\mathfrak{g}$. Then we have that*
$$(\sigma \otimes \sigma)(R) = R^{-1}, \quad (\theta \otimes \theta)(R) = R^{21}.$$

PROOF. It suffices to observe that (2.1.1), (2.1.4), (2.1.5), and (2.3.22) still hold when R is replaced by $(\sigma \otimes \sigma)(R)$ or $(\theta \otimes \theta)(R)$. Then using the uniqueness of the universal quantum R-matrix, we obtain the result of the proposition. □

DEFINITION 2.3.17. We say that a quasi-triangular (resp. triangular, resp. coboundary) Hopf algebra (A, R) is a *quantization* of a quasi-triangular (resp. triangular, resp. coboundary) Lie bialgebra $(\mathfrak{g}, r)$ if

- A is a quantization of the Lie bialgebra $\mathfrak{g}$;
- $r = h^{-1}(R - 1) \bmod h$.

EXERCISE 2.3.18. (1) Let $U_h\mathfrak{g}$ be a QUE-algebra which is a quantization of a Lie bialgebra $\mathfrak{g}$. Let $(\mathcal{D}(U_h\mathfrak{g}), R)$ be the canonical quasi-triangular structure on the double Hopf algebra. Show that it is a quantization of the double Lie bialgebra $(\mathcal{D}(\mathfrak{g}), r)$, where r is described in Proposition 2.3.6 of Chapter 1.

(2) Suppose that (A, R) is a quasi-triangular (resp. triangular, coboundary) Hopf algebra which is a quantization of a Lie bialgebra $\mathfrak{g}$. Prove that if there exists $r \in \mathfrak{g} \otimes \mathfrak{g}$ such that $R = 1 + hr + O(h^2)$, then $(\mathfrak{g}, r)$ is a quasi-triangular (resp. triangular, coboundary) Lie bialgebra.

2.4. Twisted version of $U_h\mathfrak{g}$. Let v be an invertible element of $U_h\mathfrak{g} \hat{\otimes} U_h\mathfrak{g}$ which satisfies the following properties:

(2.4.1) $$v^{12}(\Delta \otimes \mathrm{id})(v) = v^{23}(\mathrm{id} \otimes \Delta)(v),$$

(2.4.2) $$m(S \otimes \mathrm{id})(v^{12}) = m(\mathrm{id} \otimes S)(v^{12}) = 1,$$

(2.4.3) $$m(S \otimes \mathrm{id})(v^{21}) = m(\mathrm{id} \otimes S)(v^{21}) = 1.$$

This immediately implies the following proposition. The proof is a straightforward computation, which the reader may do as an exercise.

PROPOSITION 2.4.1. (1) *For any $\xi \in U_h\mathfrak{g}$, let $\Delta_v(\xi) = v\Delta(\xi)v^{-1}$. Then the deformed comultiplication Δ_v together with the same antipode S and the same counit ε (as given in (1.2.5)–(1.2.6)) define a new Hopf algebra structure on $U_h\mathfrak{g}$. The new Hopf algebra will be denoted by $U_{h,v}\mathfrak{g}$.*

(2) *$(U_{h,v}\mathfrak{g}, R_v)$ is a quasi-triangular Hopf algebra, with $R_v = v^{21}Rv^{-1}$, where R is the universal quantum R-matrix for $U_h\mathfrak{g}$.*

EXAMPLE 2.4.2. Recall that by Theorem 1.2.6, $(U\mathfrak{h})[[h]] \simeq U_h\mathfrak{h}$ can be regarded as embedded into $U_h\mathfrak{g}$. It is easy to see that $v = \exp hv_0$, where $v_0 \in \mathfrak{h} \otimes \mathfrak{h}$ satisfies (2.4.1)–(2.4.3), hence provides an example of "twisted" quantized universal enveloping algebra $U_{h,v}\mathfrak{g}$.

It is clear that the quasi-classical analog of $U_{h,v}\mathfrak{g}$ must be a quasi-triangular Lie bialgebra $(\mathfrak{g}, \varphi_v)$, where φ_v is not necessarily the standard cobracket given by

(2.1.6). Let us compute the classical r-matrix for this Lie bialgebra in the framework of Example 2.4.2.

By Proposition 2.3.13 and Theorem 1.2.6, we can write $R = 1 + r_0 h + O(h^2)$, where r_0 is given by (4.1.7). Then we get that $R_v = e^{hv_0^{21}} R e^{-hv_0} = 1 + h r_v + O(h^2)$, where

(2.4.4) $$r_v = r_0 - v_0^{12} + v_0^{21}.$$

In other words, we have proved the following proposition.

PROPOSITION 2.4.3. *If $v = \exp hv_0$, where $v_0 \in \mathfrak{h} \otimes \mathfrak{h}$, then $U_{h,v}\mathfrak{g}$ is a quantization of a quasi-triangular Lie bialgebra structure on $\mathfrak{g}$, with the classical r-matrix given by (2.4.4).*

REMARK 2.4.4. Let $\mathfrak{k}$ be a compact real form of $\mathfrak{g}$ (which is well known to be unique up to conjugation). Then the Lie bialgebra structures on $\mathfrak{g}$ which arise from the "twisting" procedure are exactly the complexifications of those on $\mathfrak{k}$ described in Section 5 of Chapter 1.

The correspondence is as follows. $U_{h,v}\mathfrak{g}$ is a quantization of the complexification of the Lie bialgebra structure on $\mathfrak{k}$ which comes from $K(1,u)$, where $u = v_0^{12} - v_0^{21} \in \mathfrak{h} \wedge \mathfrak{h}$. It is worth noting that $U_{0,v}\mathfrak{g}$—that is, the result of the "twisting" procedure applied to $U\mathfrak{g}$—is a quantization of the complexification of the Lie bialgebra structure on $\mathfrak{k}$ which comes from $K(0,u)$.

3. Center of quasi-triangular Hopf algebras

In this section we discuss important general facts about the center of quasi-triangular Hopf algebras. These results will be useful in the following chapters, while considered in the case of quantized universal enveloping algebras.

Throughout this section we deal mostly with Hopf algebras over a field $\mathbf{k}$ of characteristic zero. However, the results obtained here are applicable also to the case of a Hopf algebra over a local commutative ring $\mathfrak{R}$ complete in the $\mathfrak{m}$-adic topology, where $\mathfrak{m}$ is the maximal ideal in $\mathfrak{R}$ (deduce this from Exercise 2.1.5); in particular, they hold for the Hopf algebra $U_h\mathfrak{g}$ over $\mathbb{C}[[h]]$. We leave those as exercises to the reader.

3.1. Two central element constructions. First we give two central element constructions which can be applied to any Hopf algebra. Then we are going to specify them to the cases of almost-cocommutative and quasi-triangular Hopf algebras.

THE FIRST CENTRAL ELEMENT CONSTRUCTION. Suppose that A is a Hopf algebra over a field $\mathbf{k}$ of characteristic zero, and Z is the center of A. We can make A an $A \otimes A$-module by letting $(a_1 \otimes a_2) \circ b = a_1 b S(a_2)$. In particular, one can see that the adjoint action of A on itself can be expressed in terms of this action as follows: $a : b \mapsto \Delta(a) \circ b$.

Suppose that B is a subalgebra of $A \otimes A$ which consists of the elements $b \in A \otimes A$ commuting with $\Delta(a)$ for any $a \in A$. We define a linear map $\gamma : A \otimes A \to A$ by $\gamma(a_1 \otimes a_2) = (a_1 \otimes a_2) \circ 1 = a_1 S(a_2)$.

PROPOSITION 3.1.1. *(1) $\gamma(B) \subset Z$.*
(2) $\gamma(xb) = \gamma(x)\gamma(b)$ for any $x \in A \otimes A$, $b \in B$. In particular, the restriction of γ to B is an algebra homomorphism.

PROOF. (1) Note that $\operatorname{ad}(a)z = \Delta(a) \circ z = \varepsilon(a)z$ for any $a \in A$, $z \in Z$. Suppose $y = \gamma(b)$, where $b \in B$. Then for any $a \in A$ we have $\operatorname{ad}(a)y = \Delta(a) \circ y = \Delta(a) \circ (b \circ 1) = b \circ (\Delta(a) \circ 1) = b \circ (\varepsilon(a) \cdot 1) = \varepsilon(a)(b \circ 1) = \varepsilon(a)y$.

We show that any $y \in A$ such that $\operatorname{ad}(a)y = \varepsilon(a)y$ for any $a \in A$ lies in Z. Consider the expression $f(\Delta_3(c))$, where $c \in A$, $\Delta_3 : A \to A \otimes A \otimes A$ is the comultiplication, and $f : A \otimes A \otimes A \to A$ is given by $f(a_1 \otimes a_2 \otimes a_3) = a_1 y S(a_2) a_3$. Writing Δ_3 in the form $(\operatorname{id} \otimes \Delta)\Delta$, we get $f(\Delta_3(c)) = cy$, and writing Δ_3 in the form $(\Delta \otimes \operatorname{id})\Delta$, we get $f(\Delta_3(c)) = yc$. Thus we see that $cy = yc$, that is, $y \in Z$.

(2) We see that $\gamma(xb) = (xb) \circ 1 = x \circ \gamma(b)$. If $b \in B$, then $\gamma(b) \in Z$, hence $x \circ \gamma(b) = (x \circ 1)\gamma(b) = \gamma(x)\gamma(b)$. □

REMARK 3.1.2. Note that if $b \in B$, then $(S^{-1} \otimes S^{-1})(b) \in B$, and if $z \in Z$, then $S(z) \in Z$ and $S^{-1}(z) \in Z$. Therefore, given $\sum_k a_k \otimes b_k \in B$, we have that not only $\sum_k a_k S(b_k) \in Z$ (by Proposition 3.1.1), but also $\sum_k S^{-1}(b_k) a_k \in Z$, $\sum_k S(a_k) b_k \in Z$, and $\sum_k b_k S^{-1}(a_k) \in Z$.

THE SECOND CENTRAL ELEMENT CONSTRUCTION. Let $\mathcal{I}$ and $\bar{\mathcal{I}}$ be spaces of linear functionals $\lambda \in A^*$ on A which satisfy the conditions

(3.1.1) $\qquad \lambda(\operatorname{ad}(a)b) = \varepsilon(a)\lambda(b),$

(3.1.2) $\qquad \lambda(\operatorname{ad}'(a)b) = \varepsilon(a)\lambda(b),$

for any $a, b \in A$ respectively. Here ad' is the adjoint action of the Hopf algebra A° (the Hopf algebra with the opposite comultiplication and the skew antipode) given by

$$\operatorname{ad}'(a) : b \mapsto \sum_k a_k^{(2)} b S^{-1}(a_k^{(1)}) \quad \text{whenever} \quad \Delta(a) = \sum_k a_k^{(1)} \otimes a_k^{(2)}.$$

EXERCISE 3.1.3. Show that the conditions (3.1.1) and (3.1.2) are equivalent to

(3.1.3) $\qquad \lambda(ab) = \lambda(bS^2(a)),$

(3.1.4) $\qquad \lambda(ab) = \lambda(bS^{-2}(a)),$

respectively.

EXERCISE 3.1.4. Show that both $\mathcal{I}$ and $\bar{\mathcal{I}}$ are algebras with respect to the multiplication defined by (2.2.6). We call them algebras of ad- and ad'-invariant linear functionals on A respectively.

PROPOSITION 3.1.5. Suppose that $b \in B$. Then
(1) if $\lambda \in \mathcal{I}$, then $\zeta_b(\lambda) = (\operatorname{id} \otimes \lambda)(b) \in Z$;
(2) if $\bar{\lambda} \in \bar{\mathcal{I}}$, then $\bar{\zeta}_b(\bar{\lambda}) = (\bar{\lambda} \otimes \operatorname{id})(b) \in Z$.

PROOF. We will prove the first statement, and the second will follow from it when we replace A by A° and b by $b^{21} = \sigma(b)$, where σ is the permutation of tensor factors.

Suppose that $a \in A$ and $\Delta(a) = \sum_k a_k^{(1)} \otimes a_k^{(2)}$. Then $\sum_k \Delta(a_k^{(1)}) \otimes a_k^{(2)} = (\Delta \otimes \operatorname{id})\Delta(a) = (\operatorname{id} \otimes \Delta)\Delta(a)$, hence

$$\sum_k \left(1 \otimes S^{-1}\left(a_k^{(2)}\right)\right) \cdot \Delta\left(a_k^{(1)}\right) = a \otimes 1 = \sum_k \Delta\left(a_k^{(1)}\right) \cdot \left(1 \otimes S\left(a_k^{(2)}\right)\right).$$

Therefore,

$$\begin{aligned}
a \cdot (\mathrm{id} \otimes \lambda)(b) &= (\mathrm{id} \otimes \lambda)\left((a \otimes 1) \cdot b\right) \\
&= (\mathrm{id} \otimes \lambda)\left(\sum_k \left(1 \otimes S^{-1}\left(a_k^{(2)}\right)\right) \cdot \Delta\left(a_k^{(1)}\right) \cdot b\right) \\
&= (\mathrm{id} \otimes \lambda)\left(\sum_k \Delta\left(a_k^{(1)}\right) \cdot b \cdot \left(1 \otimes S\left(a_k^{(2)}\right)\right)\right) \\
&= (\mathrm{id} \otimes \lambda)\left(b \cdot \sum_k \Delta\left(a_k^{(1)}\right) \cdot \left(1 \otimes S\left(a_k^{(2)}\right)\right)\right) \\
&= (\mathrm{id} \otimes \lambda)\left(b \cdot (a \otimes 1)\right) = (\mathrm{id} \otimes \lambda)(b) \cdot a,
\end{aligned}$$

which implies that $(\mathrm{id} \otimes \lambda)(b) \in Z$. $\square$

REMARK 3.1.6. Note that the map $B \otimes \mathcal{I} \to Z$ given by $b \otimes \lambda \mapsto \zeta_b(\lambda) = (\mathrm{id} \otimes \lambda)(b)$ is not an algebra homomorphism in general. The reader can easily check this in the case when A is the algebra of functions on an Abelian group (so that $Z = A$, $B = A \otimes A$, and $\mathcal{I} = A^*$). The same is true for the map $B \otimes \bar{\mathcal{I}} \to Z$ given by $b \otimes \bar{\lambda} \mapsto \bar{\zeta}_b(\bar{\lambda}) = (\bar{\lambda} \otimes \mathrm{id})(b)$.

3.2. Square of the antipode in the almost-cocommutative case. An inportant fact about the almost-cocommutative Hopf algebras is that the square of the antipode turns out to be an inner automorphism. Recall that in the present section we consider the case of a Hopf algebra over a field of characteristic zero. However, the results obtained here, as well as their proofs, basically hold also for Hopf algebras over local commutative rings complete in the adic topology.

Let (A, R) be an almost-cocommutative Hopf algebra. Suppose that

$$R = \sum_k \alpha_k \otimes \beta_k, \quad R^{-1} = \sum_k \alpha^k \otimes \beta^k.$$

Consider an element

(3.2.1) $$u = \sum_k S(\beta_k)\alpha_k.$$

PROPOSITION 3.2.1. (1) *The element $u \in A$ is invertible and*

$$u^{-1} = \sum_k S^{-1}(\beta^k)\alpha^k.$$

(2) *For any $a \in A$, one has*

(3.2.2) $$S^2(a) = uau^{-1}.$$

PROOF. We will first prove that for any $a \in A$, we have

(3.2.3) $$u \cdot a = S^2(a) \cdot u.$$

Suppose that $(\Delta \otimes \mathrm{id})\Delta(a) = \sum_i a_i \otimes b_i \otimes c_i = (\mathrm{id} \otimes \Delta)\Delta(a)$. It follows from (2.1.1) that

$$(R \otimes 1) \cdot \left(\sum_i a_i \otimes b_i \otimes c_i\right) = \left(\sum_i b_i \otimes a_i \otimes c_i\right) \cdot (R \otimes 1).$$

This implies

$$\sum_{i,k} \alpha_k a_i \otimes \beta_k b_i \otimes c_i = \sum_{i,k} b_i \alpha_k \otimes a_i \beta_k \otimes c_i,$$

(3.2.4) $$\sum_{i,k} S^2(c_i) \cdot S(\beta_k b_i) \cdot \alpha_k a_i = \sum_{i,k} S^2(c_i) \cdot S(a_i \beta_k) \cdot b_i \alpha_k,$$

$$\sum_{i,k} S(b_i S(c_i)) \cdot S(\beta_k) \alpha_k a_i = \sum_{i,k} S^2(c_i) \cdot S(\beta_k) \cdot S(a_i) b_i \alpha_k.$$

It follows from the definition of antipode that $\sum_i S(a_i) b_i \otimes c_i = 1 \otimes a$ and $\sum_i a_i \otimes b_i S(c_i) = a \otimes 1$. Therefore, the left-hand side of (3.2.4) is equal to ua, while the right-hand side is equal to $S^2(a)u$. This proves (3.2.3).

Let $v = \sum_k S^{-1}(\beta^k) \alpha^k$. Then $uv = \sum_k u S^{-1}(\beta^k) \alpha^k = \sum_k S(\beta^k) u \alpha^k = \sum_{j,k} S(\beta_j \beta^k) \alpha_j \alpha^k$. Since $\sum_{j,k} \alpha_j \alpha^k \otimes \beta_j \beta^k = R \cdot R^{-1} = 1$, we have that $uv = 1$. Applying (3.2.3) with $a = v$, we get $S^2(v)u = 1$. Since u has a left inverse and a right inverse, it follows that u is invertible, hence (3.2.3) implies (3.2.2). Moreover, $u^{-1} = v$.

Note also that (3.2.2) implies that $S^2(u) = u$. $\square$

Consider the following elements:

$$u_1 = u = \sum_k S(\beta_k) \alpha_k, \qquad u_2 = \sum_k S(\alpha^k) \beta^k,$$

$$u_3 = \sum_k \beta_k S^{-1}(\alpha_k), \qquad u_4 = \sum_k \alpha^k S^{-1}(\beta^k).$$

PROPOSITION 3.2.2. (1) *The elements u_i are invertible, and*

$$u_1^{-1} = \sum_k S^{-1}(\beta^k) \alpha^k, \qquad u_2^{-1} = \sum_k S^{-1}(\alpha_k) \beta_k,$$

$$u_3^{-1} = \sum_k \beta^k S(\alpha^k), \qquad u_4^{-1} = \sum_k \alpha_k S(\beta_k).$$

(2) *For any $a \in A$ and $i = 1, 2, 3, 4$, one has $S^2(a) = u_i a u_i^{-1}$.*
(3) $S(u_1)^{-1} = u_2$, $S(u_3)^{-1} = u_4$.
(4) *The elements u_i commute with each other and the elements $u_i u_j^{-1} = u_j^{-1} u_i$ belong to the center of A.*

PROOF. By Proposition 3.2.1, we get $u_1^{-1} = u^{-1} = \sum_k S^{-1}(\beta^k) \alpha^k$. It follows that $S(u_1)^{-1} = \sum_k S(\alpha^k) \beta^k = u_2$ and $u_2^{-1} = S(u_1) = S(u) = S^{-1}(u) = \sum_k S^{-1}(\alpha_k) \beta_k$. Applying S to both sides of (3.2.2), we get $u_2 S(a) u_2^{-1} = S^3(a)$, hence $u_2 a u_2^{-1} = S^2(a)$.

Applying $S^k \otimes S^k$ to both sides of (2.1.1), we see that (2.1.1) remains true when R is replaced by $(S^k \otimes S^k)(R)$ for any $k \in \mathbf{Z}$. On replacing R by $(S^{-1} \otimes S^{-1})(R)$ the elements u_1 and u_2 are replaced by u_3 and u_4 respectively. This proves the assertions (1)–(3). The assertion (4) immediately follows from (2). $\square$

REMARK 3.2.3. We note that if A is a quasi-triangular Hopf algebra, then $(S \otimes S)(R) = R$, by Proposition 2.1.7. Therefore, in this case $u_1 = u_3$ and $u_2 = u_4$. Note also that if $R^{-1} = R^{21}$, then $u_1 = u_2$ and $u_3 = u_4$.

3.3. The quasi-triangular case.

Now let A be an almost-cocommutative Hopf algebra. Consider the above two constructions of central elements in this case. First of all, the additional structure (the quantum R-matrix) provides us with some distinguished elements of $B = \{b \in A \otimes A \mid b\Delta(a) = \Delta(a)b \text{ for any } a \in A\}$. Indeed, let R and $\tilde{R}$ be two different quantum R-matrices for A. Then it is easy to see that $\tilde{R}^{-1}R \in B$. In particular, if A is not triangular, we can put $\tilde{R} = (R^{21})^{-1}$ (see also Remark 2.1.4), where $R^{21} = \sum_k \beta_k \otimes \alpha_k$ for $R = \sum_k \alpha_k \otimes \beta_k$. Therefore, $R^{21}R \in B$.

EXERCISE 3.3.1. Let u_1, u_2, u_3, u_4 be as in Proposition 2.2.1. Show that

(3.3.1) $\quad u_1^{-1} = \gamma\left(\left(S^{-1} \otimes S^{-1}\right)\left(R^{21}\right)^{-1}\right), \quad u_2^{-1} = \gamma\left(\left(S^{-1} \otimes S^{-1}\right)(R)\right),$

(3.3.2) $\qquad u_3^{-1} = \gamma\left(\left(R^{21}\right)^{-1}\right), \quad u_4^{-1} = \gamma(R).$

REMARK 3.3.2. Replacing R by $\tilde{R} = (R^{21})^{-1}$ in (3.3.1)–(3.3.2), we get the elements $u_i u_j^{-1} \in \gamma(B)$. In particular, $u_3 u_4^{-1} = \gamma(R^{21}R)$.

Let us now assume that (A, R) is a quasi-triangular Hopf algebra. Note that in this case $u_1 = u_3 = u$ and $u_2 = u_4 = S(u)^{-1}$ (see Proposition 2.2.1). Thus the central elements $u_i u_j^{-1}$ all can be expressed in terms of the single central element

$$z = uS(u) = S(u)u.$$

Consider the element $g = uS(u)^{-1} = S(u)^{-1}u$. It is easy to see that $zg = u^2$, $zg^{-1} = S(u)^2$, $S(z) = z$, $S(g) = g^{-1}$. Moreover, for any $a \in A$, one has

(3.3.3) $\qquad\qquad gag^{-1} = S^4(a).$

The following proposition shows that g is a *group-like* element. This means that $\Delta(g) = g \otimes g$.

PROPOSITION 3.3.3. *The following formulas hold in the quasi-triangular case:*

(3.3.4) $\qquad \Delta(u) = (R^{21}R)^{-1}(u \otimes u) = (u \otimes u)(R^{21}R)^{-1},$

(3.3.5)
$$\Delta(S(u)) = (R^{21}R)^{-1}(S(u) \otimes S(u)) = (S(u) \otimes S(u))(R^{21}R)^{-1},$$

(3.3.6) $\qquad\qquad \Delta(z) = (z \otimes z)(R^{21}R)^{-2},$

(3.3.7) $\qquad\qquad \Delta(g) = g \otimes g.$

PROOF. It is easy to see that (3.3.5)–(3.3.7) follow directly from (3.3.4). Since $R^{21}R$ commutes with $\Delta(a)$ for any $a \in A$, it suffices to show that

$$\Delta(u)R^{21}R = u \otimes u.$$

In what follows, the notation R^{ij} for $R = \sum_k \alpha_k \otimes \beta_k$ means the element of a certain tensor power of A given by the formula

$$R^{ij} = \sum_k 1 \otimes \cdots \otimes 1 \otimes \alpha_k \otimes 1 \otimes \cdots \otimes 1 \otimes \beta_k \otimes 1 \otimes \cdots \otimes 1,$$

where α_k is the ith tensor factor and β_k is the jth factor. (We have written down the case when $i < j$; the case when $i > j$ can be written analogously.)

Recall that $u = \sum_k S(\beta_k)\alpha_k$. Since the antipode is an antiautomorphism of coalgebra structure (that is, $\Delta(S(a)) = (S \otimes S)(\Delta'(a))$, where $\Delta' = \sigma\Delta$ is the opposite comultiplication), we see that $\Delta(u) = \sum_k (S \otimes S)(\Delta'(\beta_k))\Delta(\alpha_k)$. We know that $R^{21}R$ commutes with $\Delta(\alpha_k)$, hence

$$\Delta(u)R^{21}R = \sum_k (S \otimes S)(\Delta'(\beta_k))R^{21}R\Delta(\alpha_k).$$

We make $A \otimes A$ a right $A \otimes A \otimes A \otimes A$-module by putting

$$(x_1 \otimes x_2) \circ (y_1 \otimes y_2 \otimes y_3 \otimes y_4) = S(y_3)x_1 y_1 \otimes S(y_4)x_2 y_2.$$

Then $\Delta(u)R^{21}R = R^{21} \circ \left(R^{12}(\Delta \otimes \Delta')(R)\right)$. By Proposition 2.1.7, we get

$$R^{12}(\Delta \otimes \Delta')(R) = R^{12}R^{13}R^{23}R^{14}R^{24} = R^{23}R^{13}R^{12}R^{14}R^{24}.$$

Also, we get

$$\begin{aligned}R^{21} \circ R^{23} &= \sum_{j,k} S(\beta_j)\beta_k \circ \alpha_k \alpha_j \\ &= (S \circ \mathrm{id})\Big(\Big(\sum_j S^{-1}(\beta_j) \circ \alpha_j\Big)\Big(\sum_k \beta_k \circ \alpha_k\Big)\Big) = 1\end{aligned}$$

since $\sum_k \alpha_k \circ \beta_k = R$ and $\sum_j \alpha_j \circ S^{-1}(\beta_j) = (\mathrm{id} \circ S^{-1})(R) = R^{-1}$. Therefore, we get $R^{21} \circ (R^{23}R^{13}) = 1 \circ R^{13} = u \otimes 1$, hence

$$R^{21} \circ R^{23}R^{13}R^{12}R^{14} = (u \otimes 1) \circ R^{12}R^{14} = \sum_{j,k} u\alpha_j \alpha_k \otimes S(\beta_k)\beta_j = u \otimes 1.$$

Finally, we see that $(u \otimes 1) \circ R^{24} = u \otimes u$ proving the assertion. $\square$

We will now discuss the second central element construction in the quasi-triangular case. Recall that we considered the algebras $\mathcal{I}$ and $\bar{\mathcal{I}}$ described by (3.1.3)–(3.1.4) (see Exercise 3.1.4). Consider also the space

(3.3.8) $\quad \mathcal{I}_0 = \{\mu \in A^* \mid \mu(ab) = \mu(ba) \text{ for any } a, b \in A\}.$

EXERCISE 3.3.4. (1) Show that $\mathcal{I}_0$ is an algebra with respect to the multiplication given by (2.2.6). Note that this is true for any almost-cocommutative Hopf algebra A. We call $\mathcal{I}_0$ the *algebra of symmetric linear functionals on A*.

(2) Prove that in the quasi-triangular case $\mathcal{I}$, $\bar{\mathcal{I}}$, and $\mathcal{I}_0$ are commutative algebras.

EXERCISE 3.3.5. (1) Let x_0 be a group-like element of a quasi-triangular Hopf algebra A (that is, $\Delta(x_0) = x_0 \otimes x_0$) such that $\mathrm{Ad}(x_0) = S^2$. Show that the vector space isomorphisms $\mathcal{I}_0 \to \mathcal{I}$ and $\mathcal{I}_0 \to \bar{\mathcal{I}}$ sending $\mu \in \mathcal{I}_0$ to $\lambda \in \mathcal{I}$ and $\bar{\lambda} \in \bar{\mathcal{I}}$ and given by

$$\lambda(a) = \mu(ax_0) \quad , \quad \bar{\lambda}(a) = \mu(ax_0^{-1})$$

are in fact algebra isomorphisms. Note that one can put $x_0 = u$.

(2) Let x be a group-like element of a quasi-triangular Hopf algebra A such that $\mathrm{Ad}(x) = S^4$. Show that the vector space isomorphism $\mathcal{I} \xrightarrow{\sim} \bar{\mathcal{I}}$ sending $\lambda \in \mathcal{I}$ into $\bar{\lambda} \in \bar{\mathcal{I}}$ and given by

$$\bar{\lambda}(a) = \lambda(ax^{-1})$$

is in fact an algebra isomorphism. Note that, by Proposition 3.3.3, for x one can take the element $g = uS(u)^{-1} = S(u)^{-1}u$.

PROPOSITION 3.3.6. *Suppose that (A, R) and $(A, \tilde{R})$ are two quasi-triangular structures on the same Hopf algebra A. If $b = \tilde{R}^{-1} R$, then ζ_b and $\bar{\zeta}_b$ are algebra homomorphisms.*

In particular, if we let $\tilde{R} = (R^{21})^{-1}$, then we get $b = R^{21} R$.

PROOF. Let us consider the case of ζ_b, since the case of $\bar{\zeta}_b$ can be proved analogously. Suppose $\lambda_1, \lambda_2 \in \mathcal{I}$. Then

$$\begin{aligned}\zeta_b(\lambda_1 \lambda_2) &= (\mathrm{id} \otimes \lambda_1 \lambda_2)(b) = (\mathrm{id} \otimes \lambda_1 \otimes \lambda_2)((\mathrm{id} \otimes \Delta)(b))\\ &= (\mathrm{id} \otimes \lambda_1 \otimes \lambda_2)\left(\left(\tilde{R}^{12}\right)^{-1} b^{13} R^{12}\right)\\ &= (\mathrm{id} \otimes \lambda_1)\left(\tilde{R}^{-1}(\zeta_b(\lambda_2) \otimes 1) R\right).\end{aligned}$$

By Proposition 3.1.5, $\zeta_b(\lambda_2) \in Z$, so that we get

$$\zeta_b(\lambda_1 \lambda_2) = (\mathrm{id} \otimes \lambda_1)\left(\tilde{R}^{-1} R(\zeta_b(\lambda_2) \otimes 1)\right) = \zeta_b(\lambda_1)\zeta_b(\lambda_2).$$

The following proposition (see also Exercise 3.3.5) shows that in the quasi-triangular case the sets of central elements obtained as $\zeta_b(\mathcal{I})$ and as $\bar{\zeta}_b(\bar{\mathcal{I}})$ are the same for $b = R^{21} R$. □

PROPOSITION 3.3.7. *Suppose that (A, R) is a quasi-triangular Hopf algebra and $b = R^{21} R$. If $\lambda \in \mathcal{I}$ and $\bar{\lambda} \in \bar{\mathcal{I}}$ are such that $\bar{\lambda}(a) = \lambda(ag^{-1})$ for any $a \in A$, where $g = uS(u)^{-1} = S(u)^{-1} u$, then $\zeta_b(\lambda) = \bar{\zeta}_b(\bar{\lambda})$.*

PROOF. By Exercise 3.3.5, we have that $\lambda(a) = \mu(au)$ for some $\mu \in \mathcal{I}_0$. Then $\bar{\lambda}(a) = \mu(aS(u))$. Recall that if $R = \sum_k \alpha_k \otimes \beta_k$, then $u = \sum_k S(\beta_k)\alpha_k$ and $S(u) = \sum_k \alpha_k S(\beta_i)$ (because of (2.1.9)). Therefore

$$\zeta_b(\lambda) = \sum_{i,j,k} \mu(\alpha_i \beta_j S(\beta_k)\alpha_k)\beta_i \alpha_j = \sum_{i,j,k} \mu(\alpha_k \alpha_i \beta_j S(\beta_k))\beta_i \alpha_j.$$

Now it suffices to prove that

$$\sum_{i,j,k} \alpha_k \alpha_i \otimes \beta_i \alpha_j \otimes \beta_j S(\beta_k) = \sum_{i,j,k} \alpha_j \alpha_k \otimes \alpha_i \beta_j \otimes S(\beta_k)\beta_i$$

or the equivalent relation

$$\sum_{i,j,k} S(\alpha_i) S(\alpha_k) \otimes \beta_i \alpha_j \otimes \beta_j S(\beta_k) = \sum_{i,j,k} S(\alpha_k) S(\alpha_j) \otimes \alpha_i \beta_j \otimes S(\beta_k)\beta_i.$$

Since $(S \otimes \mathrm{id})(R) = R^{-1}$ and $(S \otimes S)(R) = R$, this can be written in the form $\left(R^{12}\right)^{-1} R^{23} R^{13} = R^{13} R^{23} \left(R^{12}\right)^{-1}$, which is equivalent to (2.1.8). □

4. Center of $U_h \mathfrak{g}$ and quantum Harish-Chandra homomorphism

Now we return to the case of $U_h \mathfrak{g}$. From now on $R \in U_h \mathfrak{g} \hat{\otimes} U_h \mathfrak{g}$ is the universal quantum R-matrix. We know that $R \in U_h \mathfrak{g} \hat{\otimes} U_h \mathfrak{g}$ and the multiplication m in $U_h \mathfrak{g}$ is continuous in the h-adic topology. In particular, $u = m(S \otimes \mathrm{id})\left(R^{21}\right) \in U_h \mathfrak{g}$.

This allows us to show that the results of the previous section are still valid in the case of the quasi-triangular Hopf algebra $U_h \mathfrak{g}$ over $\mathbb{C}[[h]]$.

4. CENTER OF $U_h\mathfrak{g}$ AND QUANTUM HARISH-CHANDRA HOMOMORPHISM

4.1. Central elements of $U_h\mathfrak{g}$. We start with the first central element construction. As a result, we get a central element
$$z = uS(u) = S(u)u = \gamma(R^{21}R),$$
where $u = \sum_k S(\beta_k)\alpha_k$ and $R = \sum_k \alpha_k \otimes \beta_k$. However, in the case of $U_h\mathfrak{g}$ there is a much nicer central element which will be given by Corollary 4.1.2.

Indeed, that z belongs to the center follows from the fact that $S^2 = \operatorname{Ad} u_i$ with $u_1 = u$ and $u_2 = S(u)^{-1}$ (see Proposition 3.2.2). The new central element arises from a simpler formula for the square of antipode in our particular case.

Namely, suppose that we have chosen a system of simple coroots in $\mathfrak{g}$. Consider the embedding of $U\mathfrak{h}$ into $U_h\mathfrak{g}$ such that the elements H_i ($i = 1, \ldots, n$) are the images of the simple coroots. Let $\rho \in \mathfrak{h}$ be half a sum of all positive coroots, that is, an element of $\mathfrak{h}$ such that $(\rho, \omega_i) = 1$ for any $i = 1, \ldots, n$, where $\{\omega_i\}$ is the system of fundamental weights which form a basis of $\mathfrak{h}^*$ dual to the basis $\{H_i\}$ of $\mathfrak{h}$.

EXERCISE 4.1.1. Show that
$$S^2 = \operatorname{Ad}\left(e^{-h\rho}\right).$$

Hint: Check on the generators.

PROPOSITION 4.1.2. *Suppose that u, z, g are the same as in Section 3.3. Then the following properties hold:*
1. *$e^{-h\rho}u = ue^{-h\rho}$ belongs to the center of $U_h\mathfrak{g}$;*
2. *$e^{-h\rho}u = e^{h\rho}S(u)$ belongs to the center of $U_h\mathfrak{g}$;*
3. *$z = \left(e^{-h\rho}u\right)^2$ and $g = e^{2h\rho}$;*
4. *$\Delta\left(e^{-h\rho}u\right) = \left(e^{-h\rho}u \otimes e^{-h\rho}u\right)\left(R^{21}R\right)^{-1}.$*

PROOF. By Exercise 3.1.3, $u, e^{-h\rho} \in \mathcal{I}$. Since $\mathcal{I}$ is a commutative subalgebra (cf. Exercise 3.3.4), we get the statement (1) of the proposition. Moreover, $S\left(e^{-h\rho}u\right) = e^{-h\rho}u = e^{h\rho}S(u)$ is also a central element in $U_h\mathfrak{g}$. We leave the proofs of the last two statements as an exercise to the reader. As a hint, we indicate that Proposition 3.3.3 should be used. □

Now we discuss the second central element construction in the case of $U_h\mathfrak{g}$. As a result we will get some distinguished central elements. First of all, note that in this particular case a stronger version of Proposition 3.3.6 holds. Namely, we have the following proposition. (Note that there is an analogous fact for the version of the construction which deploys $\bar{\mathcal{I}}$ instead of $\mathcal{I}$.)

PROPOSITION 4.1.3. *Consider the quasi-triangular Hopf algebra $(U_h\mathfrak{g}, R)$. Let $\mathcal{I}$ be the algebra of ad_q-invariant functionals defined by (3.1.1)–(3.1.2), and Z_h the center of $U_h\mathfrak{g}$. Then the map $\zeta_\mathcal{I} : \mathcal{I} \to Z_h$ given by*

(4.1.1) $$\lambda \in \mathcal{I} \mapsto (\operatorname{id} \otimes \lambda)\left(\left(R^{21}\right)^{-1} R\right) \in Z_h$$

is an algebra isomorphism.

PROOF. Since $R \in U_h\mathfrak{g}\hat{\otimes}U_h\mathfrak{g}$, we get that $(U_h\mathfrak{g}/h^n U_h\mathfrak{g}, R \bmod h^n)$ is a quasi-triangular Hopf algebra. Applying Proposition 3.3.6 to it and passing to the limit with respect to n, we get the statement of the proposition. □

Before we consider the quantum construction, we can get an idea looking at the classical picture described in the following exercise (see also [**Dix96**]).

EXERCISE 4.1.4. (1) Show that $\mathcal{I}/h\mathcal{I}$ is the algebra of $\mathfrak{g}$-invariant polynomials on $\mathfrak{g}$. It can be identified with the algebra of $\mathfrak{g}$-invariant elements of the symmetric algebra $S(\mathfrak{g}^*)$.

(2) Show that the classical analog $Z_h/hZ_h \xrightarrow{\sim} \mathcal{I}/h\mathcal{I}$ of the isomorphism $Z_h \xrightarrow{\sim} \mathcal{I}$ given by (4.1.1) is the isomorphism $Z \xrightarrow{\sim} S(\mathfrak{g}^*)^{\mathfrak{g}}$ which comes from the projection of $U\mathfrak{g}$ onto its adjoint graded algebra $\mathrm{gr}(U\mathfrak{g}) \simeq S(\mathfrak{g}^*)$.

Now, in order to get central elements via the second construction we need some particular ad_q-invariant functionals, i.e. some elements of the commutative algebra $\mathcal{I}$. Recall that in the classical case one can consider the central elements whose images under the isomorphism $Z \xrightarrow{\sim} S(\mathfrak{g}^*)^{\mathfrak{g}}$ are the elementary symmetric polynomials on $\mathfrak{g}$ (the Harish-Chandra theorem, cf. [**Dix96**]). But the elementary symmetric polynomials on $\mathfrak{g}$ are nothing but the characters of the fundamental representations of $U\mathfrak{g}$.

The theory of finite-dimensional representations of $U_h\mathfrak{g}$ will be discussed in Section 5. We will see that the category of finite-dimensional representations of $U_h\mathfrak{g}$ and that of finite-dimensional representations of $U\mathfrak{g}$ are equivalent (see Theorem 5.1.2). In particular, we have quantum analogs of the fundamental representations. We will observe also that all the characters of finite-dimensional representations of $U_h\mathfrak{g}$ belong to the algebra $\mathcal{I}_0$ defined by (3.3.8) (see Remark 5.1.8). In view of Exercise 3.3.5, we get the following picture.

Let π_{ω_i} ($i = 1, \ldots, n = \mathrm{rank}\,\mathfrak{g}$) be the fundamental representations of $U_h\mathfrak{g}$ defined by Definition 5.1.4. Let

$$(4.1.2) \qquad C_h^{(i)} = (\mathrm{id} \otimes \lambda_i)\left(\left(R^{21}\right)^{-1} R\right),$$

where $\lambda_i(\xi) = \chi_{\pi_{\omega_i}}(\xi e^{h\rho})$, $\chi_{\pi_{\omega_i}}$ is the character of π_{ω_i}, and $\rho \in \mathfrak{h}$ is the half-sum of all positive coroots. By Proposition 3.1.5, we have that $C_h^{(i)} \in Z_h$.

DEFINITION 4.1.5. The elements $C_h^{(i)}$ defined by (4.1.2) are called *quantum Casimir elements* (or simply *quantum Casimirs*).

EXAMPLE 4.1.6. Consider the case $\mathfrak{g} = \mathfrak{sl}(2,\mathbb{C})$. Then there is only one quantum Casimir element $C_h = C_h^{(1)}$ given by

$$(4.1.3) \qquad C_h = \frac{1}{2}(X^+ X^- + X^- X^+) + \cosh\frac{h}{2}\left(\frac{\sinh hH/4}{\sinh h/2}\right)^2.$$

The classical analog is the well-known classical quadratic Casimir element

$$C = \frac{1}{2}(X^+ X^- + X^- X^+) + \frac{1}{4}H^2.$$

4.2. Quantum Harish-Chandra homomorphism.

THE CLASSICAL HARISH-CHANDRA HOMOMORPHISM. Suppose that $\mathfrak{g}$ is a finite-dimensional simple complex Lie algebra, and $\mathfrak{h}$ a Cartan subalgebra with a fixed invariant scalar product. Suppose that we have also chosen a system of Chevalley–Cartan–Weyl generators H_i, X_i^+, X_i^- ($i = 1, \ldots, n = \mathrm{rank}\,\mathfrak{g}$), so that $U\mathfrak{g}$ is generated by these generators subject to relations (2.1.4)–(2.1.5).

The classical Harish-Chandra homomorphism maps the center Z of $U\mathfrak{g}$ to $U\mathfrak{h}$. It can be described as follows. The Poincaré–Birkhoff–Witt theorem says that

$$(4.2.1) \qquad U\mathfrak{g} \xleftarrow{\sim} U\mathfrak{n}_- \otimes U\mathfrak{h} \otimes U\mathfrak{n}_+,$$

where $U\mathfrak{n}_\pm$ is the subalgebra of $U\mathfrak{g}$ generated by the elements $X_i^\pm$, $U\mathfrak{h}$ is generated by H_i, and the map in (4.2.1) is given by the multiplication in $U\mathfrak{g}$. Consider the linear map $\varepsilon_- \otimes \mathrm{id} \otimes \varepsilon_+ : U\mathfrak{g} \to U\mathfrak{h}$, where $\varepsilon_\pm$ is the restriction of the counit $\varepsilon : U\mathfrak{g} \to \mathbb{C}$ (given by $\varepsilon(1) = 1$, $\varepsilon(\mathfrak{g}) = 0$) to $U\mathfrak{n}_\pm$. It turns out that the restriction ζ of this map to Z is a homomorphism. It is called the *Harish-Chandra homomorphism*.

As is well known (see [**Dix96**]), ζ maps Z isomorphically onto the algebra $(U\mathfrak{h})^W$ of the elements of $U\mathfrak{h}$ that are invariant with respect to a certain action of the Weyl group W. This action can be described as follows. Recall that W acts naturally on $\mathfrak{h}^*$ and hence, on $\mathfrak{h}$ and on $U\mathfrak{h}$. But we will consider another action of W on $U\mathfrak{h}$ given as follows.

Let $\rho \in \mathfrak{h}^*$ be the half a sum of all positive roots. Consider the affine action of W on $\mathfrak{h}^*$ given by

$$w \cdot \alpha = w(\alpha - \rho) + \rho.$$

This induces an action of W on $U\mathfrak{h}$ given by

$$w \cdot \xi = w(\xi) - \langle \xi, \rho \rangle + \langle w(\xi), \rho \rangle,$$

where $\langle\,,\,\rangle$ is the natural pairing between $U\mathfrak{h}$ and $(U\mathfrak{h})^*$ (here we consider ρ as an element of $(U\mathfrak{h})^*$). This is the action we mean when writing $\zeta : Z \xrightarrow{\sim} (U\mathfrak{h})^W$.

Let us discuss now the quantum situation. By Theorem 1.2.6, there is an isomorphism $U_h\mathfrak{g} \xrightarrow{\sim} U\mathfrak{g}[[h]]$ which is the identity modulo h and which is the identity on $U_h\mathfrak{h} = (U\mathfrak{h})[[h]]$. In particular, we have a canonical isomorphism of the center Z_h of $U_h\mathfrak{g}$ and $Z[[h]]$.

DEFINITION 4.2.1. The composition of the canonical isomorphism $Z_h \xrightarrow{\sim} Z[[h]]$ and $\zeta \hat{\otimes} \mathbb{C}[[h]]$ is a homomorphism $\zeta_h : Z_h \to (U\mathfrak{h})[[h]]$ which is called *quantum Harish-Chandra homomorphism*.

COROLLARY 4.2.2. *The quantum Harish-Chandra homomorphism ζ_h maps Z_h isomorphically onto $(U\mathfrak{h})^W[[h]]$.*

In order to describe the quantum Harish-Chandra homomorphism, we will need some additional facts. A proof of the following theorem can be found, for example, in [**Lus93**].

THEOREM 4.2.3. *Let $U_h\mathfrak{n}_\pm$ be the subalgebra of $U_h\mathfrak{g}$ generated by the elements $X_i^\pm$, and $U_h\mathfrak{h}$ the subalgebra generated by H_i (notice that it is not a Hopf subalgebra). Then the multiplication map in $U_h\mathfrak{g}$ defines an isomorphism*

(4.2.2) $$U_h\mathfrak{g} \xleftarrow{\sim} U_h\mathfrak{n}_- \hat{\otimes} U_h\mathfrak{h} \hat{\otimes} U_h\mathfrak{n}_+.$$

Consider the linear map $\varepsilon_- \hat{\otimes} \mathrm{id} \hat{\otimes} \varepsilon_+ : U_h\mathfrak{g} \to U_h\mathfrak{h}$, where $\varepsilon_\pm$ is the restriction of the counit $\varepsilon : U_h\mathfrak{g} \to \mathbb{C}$ to $U_h\mathfrak{n}_\pm$. One can show that the restriction of $\varepsilon_- \hat{\otimes} \mathrm{id} \hat{\otimes} \varepsilon_+$ to Z_h yields a homomorphism. Recalling that $U_h\mathfrak{h}$ is canonically isomorphic to $(U\mathfrak{h})[[h]]$, we get a homomorphism $Z_h \to (U\mathfrak{h})[[h]]$ which can be shown to coincide with the quantum Harish-Chandra homomorphism.

Let us discuss now the relation between the quantum Harish-Chandra homomorphism and the second central element construction. By Proposition 4.1.3, we have an isomorphism $\zeta_\mathcal{I} : \mathcal{I} \xrightarrow{\sim} Z_h$. Let $\tilde{\zeta}_0 : \mathcal{I}_0 \to (U\mathfrak{h})[[h]]$ be the composition of the isomorphism $\mathcal{I}_0 \xrightarrow{\sim} \mathcal{I}$ given by

$$\mu \in \mathcal{I}_0 \mapsto \lambda \in \mathcal{I}, \qquad \lambda(\xi) = \mu(\xi e^{h\rho}),$$

the isomorphism $\zeta_\mathcal{I}$, and the quantum Harish-Chandra homomorphism $\zeta_h : Z_h \to (U\mathfrak{h})[[h]]$. It turns out that there is an explicit expression for the map $\tilde\zeta_0$.

PROPOSITION 4.2.4. *Let t_0 be the canonical element of $\mathfrak{h} \otimes \mathfrak{h}$ corresponding to the fixed invariant scalar product in $\mathfrak{h}$, i.e., $(t_0, a \otimes b) = (a, b)$ for any $a, b \in \mathfrak{h}$ (in particular, $t_0 = \sum_k I_k \otimes I_k$ for any orthonormal basis $\{I_k\}$ of $\mathfrak{h}$). Then for any $\mu \in \mathcal{I}_0$ we have*

$$\tilde\zeta_0(\mu) = (\mathrm{id} \otimes \mu)\left(e^{ht_0}\right). \tag{4.2.3}$$

PROOF. It follows from Proposition 2.3.13 that $(\mathrm{id} \otimes \mu)\left(R^{21}R\left(1 \otimes e^{h\rho}\right)\right)$ acts on the highest weight vector as $(\mathrm{id} \otimes \mu)\left(e^{ht_0}\left(1 \otimes e^{h\rho}\right)\right) \in U\mathfrak{h}[[h]]$. Therefore, $\tilde\zeta_0(\mu)$ is the image of $(\mathrm{id} \otimes \mu)\left(e^{ht_0}\left(1 \otimes e^{h\rho}\right)\right)$ under the action of the automorphism of $U\mathfrak{h}[[h]]$ sending $a \in \mathfrak{h}$ into $a - (a, \rho)$. This proves (4.2.3). □

5. Finite-dimensional $U_h\mathfrak{g}$-modules

Throughout this section we assume that $\mathfrak{g} = \mathfrak{g}(A)$ is a finite-dimensional simple complex Lie algebra with the standard Lie bialgebra structure (see Example 2.1.8 of Chapter 1). Recall that under these assumptions $U_h\mathfrak{g}$ is a unique quantization of this Lie bialgebra (see Theorem 1.3.2). Also, all modules are supposed to be left.

5.1. Finite-dimensional modules and highest weights.

DEFINITION 5.1.1. A *finite-dimensional $U_h\mathfrak{g}$-module* is a $U_h\mathfrak{g}$-module V which is a free $\mathbf{k}[[h]]$-module of finite type. The corresponding homomorphism $\pi : U_h\mathfrak{g} \to \mathrm{End}\,V$ is called *finite-dimensional representation* of $U_h\mathfrak{g}$.

The finite-dimensional $U_h\mathfrak{g}$-modules form a category which we denote by $\mathcal{M}$. Denote by $\mathfrak{g}$-mod the category of finite-dimensional $\mathfrak{g}$-modules. By Theorem 1.2.6, there exists an isomorphism $\Psi : U_h\mathfrak{g} \tilde\to (U\mathfrak{g})[[h]]$ which is the identity modulo h. Recall that it can be chosen so that it would be the identity on $\mathfrak{h}$.

Assume that we have fixed such an isomorphism. Then it defines a functor $\mathcal{F} : \mathfrak{g}\text{-mod} \to \mathcal{M}$ so that $\mathcal{F}(V) = V[[h]]$, with the action of $U_h\mathfrak{g}$ given by the composition of the action of $U\mathfrak{g}[[h]]$ with Ψ^{-1}. We also have a functor $\mathcal{G} : \mathcal{M} \to \mathfrak{g}\text{-mod}$ such that $\mathcal{G}(W) = W/hW$.

THEOREM 5.1.2. *The functors $\mathcal{F}$ and $\mathcal{G}$ define bijections between the classes of isomorphisms of objects in the categories $\mathfrak{g}$-mod and $\mathcal{M}$. The simple modules in $\mathfrak{g}$-mod correspond to the indecomposable modules in $\mathcal{M}$.*

PROOF. Straightforward. □

Theorem 5.1.2 shows that the entire classical Cartan theory of finite-dimensional $U\mathfrak{g}$-modules can be translated into the language of finite-dimensional $U_h\mathfrak{g}$-modules.

We introduce the following notation. Let $P \subset \mathfrak{g}^*$ be the weight lattice of $\mathfrak{g}$, $P_+ \subset P$ (resp. $P_- \subset P$) the set of dominant (resp. antidominant) weights of $\mathfrak{g}$, and $P_{++} \subset P_+$ (resp. $P_{--} \subset P_-$) the set of regular dominant (resp. regular antidominant) weights of $\mathfrak{g}$. As is well known, the finite-dimensional indecomposable $U\mathfrak{g}$-modules are parameterized by P_+ (as well as by $P_- = -P_+$).

Denote by $L_0(\Lambda)$ the $U\mathfrak{g}$-module corresponding to $\Lambda \in P_+$ and by $L(\Lambda)$ the corresponding $U_h\mathfrak{g}$-module.

DEFINITION 5.1.3. We will say that the indecomposable $U_h\mathfrak{g}$-module $L(\Lambda)$ is of the highest weight Λ.

By Theorem 5.1.2, the finite-dimensional indecomposable $U_h\mathfrak{g}$-modules are parameterized by the highest weights of the corresponding $U\mathfrak{g}$-modules. Denote by $L(\Lambda)$ the $U_h\mathfrak{g}$-module such that $L_0(\Lambda) = L(\Lambda)/hL(\Lambda)$ is the $U\mathfrak{g}$-module with the highest weight $\Lambda \in P_+$. The corresponding representation will be denoted by π_Λ.

DEFINITION 5.1.4. If $\Lambda = \omega_i$ is the ith fundamental weight of $\mathfrak{g}$ (that is, such that $\omega_i(H_j) = \delta_{ij}$), then the representation π_{ω_i} is called the *ith fundamental representation* of $U_h\mathfrak{g}$.

There is a weight decomposition of $L(\Lambda)$ into the eigenspaces with respect to the action of the commutative subalgebra $U_h\mathfrak{h} \subset U_h\mathfrak{g}$ generated by H_i, $i = 1, 2, \ldots, n$.

PROPOSITION 5.1.5. *Suppose that Λ is a dominant weight. Let $L_0(\Lambda)$ be the simple $U\mathfrak{g}$-module with the highest weight Λ, and $L(\Lambda) = L_0(\Lambda)[[h]]$ the corresponding indecomposable $U_h\mathfrak{g}$-module. Consider the weight decomposition of $L_0(\Lambda)$:*

$$L_0(\Lambda) = \bigoplus_{\lambda \in P(\Lambda)} L_0(\Lambda)_\lambda,$$

where $P(\Lambda)$ denotes the set of weights of $L_0(\Lambda)$, and

$$L_0(\Lambda)_\lambda = \{v \in L_0(\Lambda) \mid av = \lambda(a)v \text{ for any } a \in \mathfrak{h}\}.$$

Then we have the following weight decomposition of $L(\Lambda)$:

(5.1.1) $$L(\Lambda) = \bigoplus_{\lambda \in P(\Lambda)} L(\Lambda)_\lambda,$$

where

(5.1.2) $$L(\Lambda)_\lambda \overset{\text{def}}{=} L_0(\Lambda)_\lambda[[h]]$$
$$= \{v \in L(\Lambda) \mid av = \lambda(a)v \text{ for any } a \in \mathfrak{h} \subset U_h\mathfrak{g}\}.$$

(The last equality above is due to the fact that the isomorphism $U_h\mathfrak{g} \overset{\sim}{\to} (U\mathfrak{g})[[h]]$ is the identity on $\mathfrak{h}$.) In particular, we have the following description of $L(\Lambda)$.

COROLLARY 5.1.6. (1) *The finite-dimensional indecomposable module $L(\Lambda)$ over $U_h\mathfrak{g}$ is generated by a vector v_Λ such that*

(5.1.3) $$X_i^+ v_\Lambda = 0, \quad H_i v_\Lambda = \Lambda(H_i) v_\Lambda.$$

The vector v_Λ is called highest weight vector *of $L(\Lambda)$, and the weight Λ is called* highest weight *of $L(\Lambda)$. The highest weight vector is unique up to a scalar factor.*

(2) *The finite-dimensional indecomposable $U_h\mathfrak{g}$-module $L(\Lambda)$ is generated by a vector $v_{\tilde{\Lambda}}$ such that*

(5.1.4) $$X_i^- v_{\tilde{\Lambda}} = 0, \quad H_i v_{\tilde{\Lambda}} = \tilde{\Lambda}(H_i) v_{\tilde{\Lambda}},$$

where $\tilde{\Lambda} = w_0 \Lambda$, $w_0 \in W$ is the element of the maximal length of the Weyl group. The vector $v_{\tilde{\Lambda}}$ is called lowest weight vector *of $L(\Lambda)$, and the weight $\tilde{\Lambda}$ is called* lowest weight *of $L(\Lambda)$. The lowest weight vector is unique up to a scalar factor.*

We are not going to review here the translation of the entire classical theory to the quantum case. The reader may think of this as a series of simple exercises. However, we will give a definition of two very important notions: the notion of a matrix element and the notion of the character of a representation of $U_h\mathfrak{g}$; we regard them as linear functionals on $U_h\mathfrak{g}$.

DEFINITION 5.1.7. Let V be a finite-dimensional $U_h\mathfrak{g}$-module, and π the corresponding representation.
(1) Suppose that $v \in V$ and $l \in V^*$. The linear functional $c_{l,v}^\pi$ on $U_h\mathfrak{g}$ given by

$$(5.1.5) \qquad c_{l,v}^\pi(\xi) = l(\pi(\xi)v) \quad (\xi \in U_h\mathfrak{g})$$

is called a *matrix element* of the representation π corresponding to the pair $(l,v) \in V^* \oplus V$.

If $V = L(\Lambda)$, we denote $c_{l,v}^\pi$ also by $c_{l,v}^\Lambda$.

(2) The linear functional χ_π on $U_h\mathfrak{g}$ given by

$$(5.1.6) \qquad \chi_\pi(\xi) = \mathrm{tr}_V \pi(\xi) \quad (\xi \in U_h\mathfrak{g}),$$

where tr_V is the trace of a linear operator in V is called the *character* of the representation π.

REMARK 5.1.8. Note that for any pair of dual bases $\{v_k\}$ and $\{l_k\}$ of V and V^* respectively, one has

$$\chi_\pi = \sum_k c_{kk}^\pi, \quad \text{where} \quad c_{kk}^\pi = c_{l_k,v_k}^\pi.$$

REMARK 5.1.9. Consider the algebra $\mathcal{I} \subset (U_h\mathfrak{g})^*$ of symmetric linear functionals on $U_h\mathfrak{g}$ defined by (3.3.8). By the property of the trace that $\mathrm{tr}_V(ab) = \mathrm{tr}_V(ba)$, we see that the characters of finite-dimensional representations of $U_h\mathfrak{g}$ belong to $\mathcal{I}_0$. In particular, by Exercise 3.3.4, they commute.

EXAMPLE 5.1.10. Consider the case $\mathfrak{g} = \mathfrak{sl}(2,\mathbb{C})$. The finite-dimensional indecomposable $U_h\mathfrak{sl}(2,\mathbb{C})$-modules are parameterized, as in the classical case, by the discrete parameter $m = 0, 1, \ldots$. Let us denote the corresponding modules by V_m and the corresponding representations by ρ_m. The rank of V_m over $\mathbb{C}[[h]]$ is equal to $m+1$. The fundamental representation of $U_h\mathfrak{sl}(2,\mathbb{C})$ corresponds to $m = 1$.

EXERCISE 5.1.11. Show that one can choose a basis $\{e_k\}$ in V_m such that the finite-dimensional representation ρ_m of $U_h\mathfrak{sl}(2,\mathbb{C})$ is given as follows:

$$(5.1.7) \qquad \rho_m(X): e_k \mapsto \frac{q^{k-m} - q^{-k+m}}{q - q^{-1}} e_{k+1},$$

$$(5.1.8) \qquad \rho_m(Y): e_k \mapsto \frac{q^k - q^{-k}}{q - q^{-1}} e_{k-1},$$

$$(5.1.9) \qquad \rho_m(H): e_k \mapsto (2k - m)e_k,$$

where $k = 0, 1, \ldots, m$. Here the highest weight vector is e_m and the lowest weight vector is e_0.

REMARK 5.1.12. It is easy to see that the basis $\{e_k\}$ consists of eigenvectors of $\rho_m(H)$. Since each eigenspace of $\pi_l(H)$ is one-dimensional, the basis $\{e_k\}$ is fixed by this condition up to multiplication by a constant.

5.2. Central characters and the quantum Harish-Chandra homomorphism.

Recall that by the Schur lemma, any central element of $U_h\mathfrak{g}$ must act as multiplication by a scalar in any irreducible finite-dimensional representation of $U_h\mathfrak{g}$.

DEFINITION 5.2.1. Let π be an irreducible finite-dimensional representation of $U_h\mathfrak{g}$. The *central character* of π is a homomorphism $c_\pi : Z_h \to \mathbb{C}[[h]]$ given by

$$(5.2.1) \qquad c_\pi : \xi \mapsto \pi(\xi) \in \mathbb{C}[[h]]$$

(we will use also the notation c_V, where V is the module corresponding to the representation π, and a shortened notation $c_\Lambda = c_{L(\Lambda)}$ if $V = L(\Lambda)$).

In the classical case there is a simple relation between the central characters and the Harish-Chandra homomorphism. Namely, if ξ is a central element of $U\mathfrak{g}$ and $f \in (U\mathfrak{h})^W$ is the image of ξ under the Harish-Chandra homomorphism, then for any simple finite-dimensional $U\mathfrak{g}$-module V_0, ξ acts on V_0 as multiplication by $f(\Lambda+\rho) = f(\tilde\Lambda-\rho)$, where Λ and $\tilde\Lambda = w_0\Lambda$ are the highest and the lowest weights of V_0 respectively. Using the fact that the isomorphism $U_h\mathfrak{g} \xrightarrow{\sim} (U\mathfrak{g})[[h]]$ is the identity on $\mathfrak{h}$, we get the corresponding statement in the quantum case.

PROPOSITION 5.2.2. *If $\xi \in Z_h$ and $f = \zeta_h(\xi) \in (U\mathfrak{h})^W[[h]]$, then for any finite-dimensional indecomposable $U_h\mathfrak{g}$-module V one has*

$$c_V(\xi) = f(\Lambda + \rho) = f(\tilde\Lambda - \rho),$$

where Λ and $\tilde\Lambda = w_0\Lambda$ are the highest and the lowest weights of V respectively.

We can use this fact in order to compute the quantum Harish-Chandra homomorphism on the central element $e^{-h\rho}u$ introduced in the previous section (see Proposition 4.1.2).

PROPOSITION 5.2.3. *Suppose that $V = L(\Lambda)$ and denote c_V by c_Λ. Then*

$$(5.2.2) \qquad c_\Lambda\left(e^{-h\rho}u\right) = \exp\left(-\frac{h}{2}(\Lambda, \Lambda + 2\rho)\right).$$

PROOF. It suffices to apply $e^{-h\rho}u$ to the highest weight vector $v_\Lambda \in V$. Recall that $u = \sum_k S(\beta_k)\alpha_k$, where $R = \sum_k \alpha_k \otimes \beta_k$. It follows from (2.3.22) that all the summands except for the first in the corresponding expression for $e^{-h\rho}uv_\Lambda$ disappear, leaving only the term with the elements from $U_h\mathfrak{h}$ acting on v_Λ. Then it is a straightforward computation to check that it is given by (5.2.2). □

COROLLARY 5.2.4. *The image $\zeta_h\left(e^{-h\rho}u\right) \in (U\mathfrak{h})^W[[h]]$ of $e^{-h\rho}u$ under the quantum Harish-Chandra homomorphism, regarded as a function of $\Lambda \in \mathfrak{h}^*$, is equal to $\exp\left(\frac{h}{2}((\rho,\rho) - (\Lambda,\Lambda))\right)$.*

The following consequence of this result shows that $e^{-h\rho}u$ can be thought of as a quantum analog of the quadratic Casimir element $C \in Z$ (another quantum analog is the quantum Casimir element $C_h = C_h^{(1)}$ introduced in Definition 4.1.5).

PROPOSITION 5.2.5. *The image of the central element*

$$-2h^{-1}\ln\left(e^{-h\rho}u\right) = -h^{-1}\ln(uS(u))$$

under the canonical isomorphism $Z_h \tilde{\to} Z[[h]]$ of the centers of $U_h\mathfrak{g}$ and $(U\mathfrak{g})[[h]]$ is the quadratic Casimir element

$$C = \sum_\alpha e_{-\alpha}e_\alpha + \sum_k I_k^2 = \sum_{\alpha>0} e_{-\alpha}e_\alpha + \sum_k I_k^2 + 2\rho,$$

where $e_\alpha \in \mathfrak{g}$ are root vectors normalized by the condition $(e_\alpha, e_{-\alpha}) = 1$, and $\{I_k\}$ is an orthonormal basis of $\mathfrak{h}$.

PROOF. It suffices to check that $-2h^{-1}\ln(e^{-h\rho}u) \in U_h\mathfrak{g} \simeq U\mathfrak{g}[[h]]$ acts on the highest weight vector in $L(\Lambda)$ as the multiplication by $(\Lambda, \Lambda + 2\rho)$. But this immediately follows from Proposition 5.2.3. □

6. Tensor products of $U_h\mathfrak{g}$-modules and tensor categories

Let A be a Hopf algebra over a commutative ring **k**. We say that an A-module M is finite-dimensional if it is projective of finite rank over **k**. There is a natural tensor product on the category of finite-dimensional A-modules which we will understand in the topological sense if **k** is a topological ring. Let us recall the definition.

DEFINITION 6.0.1. Let V_1 and V_2 be two finite-dimensional A-modules. The finite-dimensional A-module isomorphic as a **k**-module to $V_1 \otimes V_2$ and equipped with the A-action

(6.0.1) $$a : v_1 \otimes v_2 \mapsto \Delta(a)(v_1 \otimes v_2)$$

is called the *tensor product* of the A-modules V_1 and V_2 and denoted by $V_1 \otimes V_2$.

If π_i is the representation corresponding to V_i, then the representation corresponding to $V_1 \otimes V_2$ is called the *tensor product* of the representations π_1 and π_2 and denoted by $\pi_1 \otimes \pi_2$.

As we will see, the category of finite-dimensional $U_h\mathfrak{g}$-modules equipped with the tensor product yields an example of what is called a ribbon category. But first we need some definitions.

DEFINITION 6.0.2. Consider a triple $\mathcal{C} = (\tilde{\mathcal{C}}, \otimes, a)$, where $\tilde{\mathcal{C}}$ is a category, $\otimes : \tilde{\mathcal{C}} \times \tilde{\mathcal{C}} \to \tilde{\mathcal{C}}$ a functor, and $a = \{a_{XYZ}\}$ a natural transformation between the functors $\otimes(\otimes \times \text{id})$ and $\otimes(\text{id} \times \otimes)$, i.e., a is a collection of morphisms

$$a_{XYZ} : (X \otimes Y) \otimes Z \to X \otimes (Y \otimes Z)$$

for any $X, Y, Z \in \text{Ob}\,\tilde{\mathcal{C}}$.

Then $\mathcal{C}$ is called a *monoidal category*, if the following conditions are satisfied:
(1) For any $X, Y, Z \in \text{Ob}\,\tilde{\mathcal{C}}$, a_{XYZ} is an isomoprhism.
(2) (The pentagon axiom) For any $X, Y, Z, W \in \text{Ob}\,\tilde{\mathcal{C}}$, the following diagram commutes:

$$\begin{array}{ccc}
((X \otimes Y) \otimes Z) \otimes W & \xrightarrow{a_{XYZ}\otimes \text{id}} & (X \otimes (Y \otimes Z)) \otimes W \\
{\scriptstyle a_{X\otimes Y,Z,W}}\downarrow & & \downarrow{\scriptstyle a_{X,Y\otimes Z,W}} \\
(X \otimes Y) \otimes (Z \otimes W) & & X \otimes ((Y \otimes Z) \otimes W) \\
& {\scriptstyle a_{X,Y,Z\otimes W}}\searrow \quad \swarrow{\scriptstyle \text{id}\otimes a_{YZW}} & \\
& X \otimes (Y \otimes (Z \otimes W)) &
\end{array}$$

(3) There exists an object I in $\tilde{\mathcal{C}}$ such that $I \otimes I = I$, and the functors $X \mapsto X \otimes I$ and $X \mapsto I \otimes X$ are autoequivalences of $\tilde{\mathcal{C}}$.

The natural transformation a is called the *associativity constraint*.

EXAMPLE 6.0.3. Clearly the category of finite-dimensional vector spaces with the usual tensor product is a monoidal category, the associativity constraint $a_{X,Y,Z}$ given by $(x \otimes y) \otimes z \mapsto x \otimes (y \otimes z)$ for any $x \in X, y \in Y, z \in Z$ (that is, the identity map).

DEFINITION 6.0.4. A monoidal category $\mathcal{C}$ is said to be *strict* if for any objects X, Y, Z, we have

(6.0.2) $$(X \otimes Y) \otimes Z = X \otimes (Y \otimes Z),$$
(6.0.3) $$a_{X,Y,Z} = \mathrm{id},$$

and there exists an object $\mathbf{1}$ in $\tilde{\mathcal{C}}$ such that

$$X \otimes \mathbf{1} = \mathbf{1} \otimes X = X$$

for any object X in $\mathcal{C}$ such that the following diagram commutes for any objects X, Y in $\mathcal{C}$:

$$\begin{array}{ccc} X \otimes Y & = & (X \otimes \mathbf{1}) \otimes Y \\ \mathrm{id}_{X \otimes Y} \downarrow & & \downarrow a_{X,\mathbf{1},Y} \\ X \otimes Y & = & X \otimes (\mathbf{1} \otimes Y) \end{array}$$

EXAMPLE 6.0.5. It is clear that the category of finite-dimensional vector spaces is a strict monoidal category. In this case the object $\mathbf{1}$ is the one-dimensional vector space.

DEFINITION 6.0.6. We say that an object X in a strict monoidal category $\mathcal{C}$ is *rigid* if there exists an object Y in $\mathcal{C}$ and a pair of morphisms

$$i_X : \mathbf{1} \to X \otimes Y, \quad e_X : Y \otimes X \to \mathbf{1}$$

such that the following diagrams commute:

$$\begin{array}{ccc} X & = & \mathbf{1} \otimes X \\ \mathrm{id}_X \downarrow & & \downarrow i_X \otimes \mathrm{id}_X \\ X & \xleftarrow{\mathrm{id}_X \otimes e_X} & X \otimes Y \otimes X \end{array}$$

$$\begin{array}{ccc} Y & = & Y \otimes \mathbf{1} \\ \mathrm{id}_Y \downarrow & & \downarrow \mathrm{id}_Y \otimes i_X \\ Y & \xleftarrow{e_X \otimes \mathrm{id}_Y} & Y \otimes X \otimes Y \end{array}$$

In this case we say that Y is *dual* to X and denote it by $Y = X^*$.

DEFINITION 6.0.7. We say that a strict monidal category $\mathcal{C}$ is *rigid* if any object X in $\mathcal{C}$ is rigid and isomorphic to the dual object Y^* for some object Y in $\mathcal{C}$.

EXAMPLE 6.0.8. It is easy to see that the category of finite-dimensional vector spaces is rigid. Indeed, the dual object to any such vector space V is just the space of all linear functionals on V.

DEFINITION 6.0.9. Let $\mathcal{C}$ be a strict monoidal category. A *braiding* (or a *commutativity constraint*) $c = \{c_{X,Y}\}$ is a functorial isomorphism between the functors $(X, Y) \mapsto X \otimes Y$ and $(X, Y) \mapsto Y \otimes X$ in $\mathcal{C} \times \mathcal{C}$ such that
$$c_{X,\mathbf{1}} = c_{\mathbf{1},X} = \mathrm{id}_X$$
for any object X in $\mathcal{C}$, and the following diagrams commute for any objects X, Y, Z in $\mathcal{C}$ (the so-called hexagon identities):

$$
\begin{array}{ccccc}
(X \otimes Y) \otimes Z & \xrightarrow{\mathrm{id}} & X \otimes (Y \otimes Z) & \xrightarrow{\mathrm{id}_X \otimes c_{Y,Z}} & X \otimes (Z \otimes Y) \\
{\scriptstyle c_{X \otimes Y, Z}} \downarrow & & & & \downarrow {\scriptstyle \mathrm{id}} \\
Z \otimes (X \otimes Y) & \xleftarrow{\mathrm{id}} & (Z \otimes X) \otimes Y & \xleftarrow{c_{X,Z} \otimes \mathrm{id}_Y} & (X \otimes Z) \otimes Y
\end{array}
$$

$$
\begin{array}{ccccc}
X \otimes (Y \otimes Z) & \xrightarrow{\mathrm{id}} & (X \otimes Y) \otimes Z & \xrightarrow{c_{X,Y} \otimes \mathrm{id}_Z} & (Y \otimes X) \otimes Z \\
{\scriptstyle c_{X, Y \otimes Z}} \downarrow & & & & \downarrow {\scriptstyle \mathrm{id}} \\
(Y \otimes Z) \otimes X & \xleftarrow{\mathrm{id}} & Y \otimes (Z \otimes X) & \xleftarrow{\mathrm{id}_Y \otimes c_{X,Z}} & Y \otimes (X \otimes Z)
\end{array}
$$

DEFINITION 6.0.10. A *braided strict monoidal category* is a strict monoidal category equipped with a braiding. It is called *rigid* if it is also a rigid strict monoidal category.

EXAMPLE 6.0.11. The category of finite-dimensional vector spaces can be equipped with a braided category structure by taking the permutation $c_{X,Y} : x \otimes y \mapsto y \otimes x$ for $x \in X$, $y \in Y$.

DEFINITION 6.0.12. Let $\mathcal{C}$ be a braided strict monoidal category. A *balancing* is an automorphism $b = \{b_X\}$ of the identity functor such that the following diagram commutes for any objects X, Y in $\mathcal{C}$:

$$
\begin{array}{ccc}
X \otimes Y & \xrightarrow{b_{X \otimes Y}} & X \otimes Y \\
{\scriptstyle b_X \otimes b_Y} \downarrow & & \downarrow {\scriptstyle c_{X,Y}} \\
X \otimes Y & \xleftarrow{c_{Y,X}} & Y \otimes X
\end{array}
$$

DEFINITION 6.0.13. A *balanced braided monoidal category* is a braided strict monoidal category equipped with a balancing. A rigid balanced braided monoidal category is called a *ribbon category*.

EXAMPLE 6.0.14. It is easy to see that the braided category of finite-dimensional vector spaces defined in Example 6.0.11 is in fact a ribbon category, the balancing being given by the identity maps.

THEOREM 6.0.15. *The category $\mathcal{M}$ of finite-dimensional $U_h\mathfrak{g}$-modules is a ribbon category.*

PROOF. We define the associativity and commutativity constraints, as well as balancing and duality functor on $\mathcal{M}$ and prove that they give rise to a ribbon category structure.

There is an obvious forgetful functor from $\mathcal{M}$ to the category of free modules of finite rank over $\mathbb{C}[[h]]$. Define the associativity constraint a_{XYZ} to be the identity map of the corresponding complex vector spaces. It is easy to see that this defines a monoidal structure on $\mathcal{M}$. The fact that the pentagon identity is satisfied follows from the coassociativity of the comultiplication in $U_h\mathfrak{g}$.

Define the object $\mathbf{1}$ in $\mathcal{M}$ as the trivial $U_h\mathfrak{g}$-module corresponding to the counit $\varepsilon : U_h\mathfrak{g} \to \mathbb{C}[[h]]$. This yields a strict monoidal category structure on $\mathcal{M}$ if we use the standard identification of $\mathbb{C}[[h]]$-modules $(M \otimes N) \otimes P = M \otimes (N \otimes P)$ by $(a \otimes b) \otimes c = a \otimes (b \otimes c)$ for any $a \in M, b \in N, c \in P$.

Moreover, it is easy to check that the usual definition of the dual $U_h\mathfrak{g}$-module gives the rigid strict monoidal category on $\mathcal{M}$:

$$X^* = \operatorname{Hom}_{\mathbb{C}[[h]]}^{\operatorname{cont}}(X, \mathbb{C}[[h]]),$$

the action of $U_h\mathfrak{g}$ given by

$$a : l \mapsto S(a)^*(l),$$

where $a \in U_h\mathfrak{g}$, $l \in X^*$, S is the antipode in $U_h\mathfrak{g}$, and $*$ indicates the dual linear map. The notation $\operatorname{Hom}^{\operatorname{cont}}$ stands for the space of *continuous* linear maps. The fact that any object X in $\mathcal{M}$ is dual to some other object Y follows from the invertibility of the antipode S.

Let $R \in U_h\mathfrak{g}\hat{\otimes}U_h\mathfrak{g}$ be the universal quantum R-matrix. Then for any finite-dimensional $U_h\mathfrak{g}$-modules X and Y, we define $c_{XY} : X \otimes Y \to Y \otimes X$ by

(6.0.4) $$c_{XY} = P_{XY} R_{XY},$$

where $R_{XY} : X \otimes Y \to Y \otimes X$ is the action of R on $X \otimes Y$, and $P_{XY} : X \otimes Y \to Y \otimes X$ is the permutation $x \otimes y \mapsto y \otimes x$ (for any $x \in X, y \in Y$). It follows from (2.1.1) that (6.0.4) is a braiding on $\mathcal{M}$. Therefore, $\mathcal{M}$ is a rigid braided strict monoidal category.

Finally, let $b_X : X \to X$ be the action of the element $e^{-h\rho}u$, where u is defined by (3.2.1). As follows from Proposition 4.1.2 (4), b_X is a balancing on $\mathcal{M}$. Therefore, $\mathcal{M}$ is a ribbon category. $\square$

7. Fixed quantization parameter

Although the quantization problem is stated in the framework of the topological Hopf algebras over $\mathbf{k}[[h]]$, in certain cases one can define an analog of the quantized algebra over $\mathbf{k}$. This would, roughly speaking, correspond to "fixing" the quantization parameter h.

This can be done, for instance, in the case when we quantize a complex simple Lie algebra $\mathfrak{g}$ with the standard Lie bialgebra structure. Recall that the quantized algebra in this case is nothing but $U_h\mathfrak{g}$ defined in Section 1.2 of this chapter. In this section we define its analog over the field $\mathbb{C}$ of complex numbers and briefly sketch some of the properties. The proofs and a more detailed exposition can be found in [**Lus93**].

7.1. The complex Hopf algebra $U_q\mathfrak{g}$.

DEFINITION 7.1.1. Let $\mathfrak{g}$ be a complex simple Lie algebra equipped with the standard Lie bialgebra structure, $q \in \mathbb{C}\backslash\{0\}$ a nonzero complex number which is not a root of unity. Let $A = (a_{ij})_{i,j=1}^n$ be the Cartan matrix of $\mathfrak{g}$, where $a_{ij} = 2\frac{(\alpha_i,\alpha_j)}{(\alpha_j,\alpha_j)}$ and α_i $(i = 1, 2, \ldots, n)$ are the simple roots.

The *quantum universal enveloping algebra* $U_q\mathfrak{g}$ is the unital complex algebra with the generators X_i^+, X_i^-, K_i $(i = 1, 2, \ldots, n)$ such that K_i is invertible for any i and the following relations are satisfied for any $i, j = 1, 2, \ldots, n$:

$$(7.1.1) \qquad K_i K_j = K_j K_i, \quad K_i X_j^\pm = q^{\pm(\alpha_i,\alpha_j)} X_j^\pm K_i,$$

$$(7.1.2) \qquad X_i^+ X_j^- - X_j^- X_i^+ = \delta_{ij} \frac{K_i^2 - K_i^{-2}}{q_i - q_i^{-1}},$$

$$(7.1.3) \qquad \sum_{k=0}^{1-a_{ij}} (-1)^k \binom{1-a_{ij}}{k}_{q_i} (X_i^\pm)^k X_j^\pm (X_i^\pm)^{1-a_{ij}-k} = 0 \quad \text{for } i \neq j,$$

where $q_i = q^{\frac{(\alpha_i,\alpha_i)}{2}}$. Here $\binom{m}{n}_t = \frac{(t;t)_m}{(t;t)_n(t;t)_{m-n}}$ is the so-called t-binomial coefficient (or Gauss polynomial), $(a;t)_k = (1-a)(1-at)\cdots(1-at^{k-1})$.

PROPOSITION 7.1.2. *The quantum universal enveloping algebra $U_q\mathfrak{g}$ carries a Hopf algebra structure with the comultiplication Δ, the antipode S, and the counit ε defined on the generators by*

$$\Delta(K_i) = K_i \otimes K_i, \qquad \Delta\left(X_i^\pm\right) = X_i^\pm \otimes K_i + K_i^{-1} \otimes X_i^\pm,$$
$$S(K_i) = K_i^{-1}, \qquad S\left(X_i^\pm\right) = -q_i^{\pm 1} X_i^\pm,$$
$$\varepsilon(K_i) = 1, \qquad \varepsilon\left(X_i^\pm\right) = 0.$$

REMARK 7.1.3. Note that $U_q\mathfrak{g}$ can be defined as a complex Hopf subalgebra of $U_h\mathfrak{g}$.

7.2. Quasi-R-matrix.

Although $U_q\mathfrak{g}$ does resemble $U_h\mathfrak{g}$, there are several differences between the two. For example, the complex Hopf algebra $U_q\mathfrak{g}$ is not quasi-triangular. One of the reasons is that the element q^{t_0}, where t_0 is the canonical element in $\mathfrak{h} \otimes \mathfrak{h}$ (cf. (2.3.22)), cannot be expressed in terms of the generators K_i $(i = 1, 2, \ldots, n)$.

However, one can define a similar element Θ called *quasi-R-matrix* in [**Lus93**]. This will be done in Theorem 7.2.3 below.

DEFINITION 7.2.1. Define $U_q\mathfrak{b}_+$ (resp. $U_q\mathfrak{b}_-$) as the Hopf subalgebra of $U_q\mathfrak{g}$ generated by $X_i^+, K_i^{\pm 1}$ (resp. by $X_i^-, K_i^\pm$) $(i = 1, 2, \ldots, n)$. Also, define $U_q\mathfrak{h}$ as the Hopf subalgebra of $U_q\mathfrak{g}$ generated by $K_i^\pm$ $(i = 1, 2, \ldots, n)$. Finally, define $U_q\mathfrak{n}^\pm$ as the subalgebra (not a Hopf subalgebra) of $U_q\mathfrak{g}$ generated by $X_i^\pm$ $(i = 1, 2, \ldots, n)$.

Consider the grading on $U_q\mathfrak{n}_\pm$ given by

$$U_q\mathfrak{n}_\pm = \bigoplus_{\nu=0}^\infty (U_q\mathfrak{n}_\pm)_\nu,$$

where $(U_q\mathfrak{n}_\pm)_\nu$ is the subspace of $U_q\mathfrak{n}_\pm$ spanned by monomials $X_{i_1}^\pm X_{i_2}^\pm \cdots X_{i_\nu}^\pm$ of length ν. It induces a filtration

$$F_N U_q\mathfrak{n}_\pm = \bigoplus_{\nu=N}^{\infty} (U_q\mathfrak{n}_\pm)_\nu.$$

Following [**Lus93**], we introduce the notation

$$\operatorname{tr}\nu = i_1 + i_2 + \cdots + i_\nu, \qquad |\nu| = \alpha_{i_1} + \alpha_{i_2} + \cdots + \alpha_{i_\nu},$$

where α_i is the root corresponding to X_i.

Then $U_q\mathfrak{g} \otimes U_q\mathfrak{g}$ carries a topology with the base consisting of the vector spaces

$$W_N = \left(U_q\mathfrak{b}^-\right)(F_N U_q\mathfrak{n}_+) \otimes U_q\mathfrak{g} + U_q\mathfrak{g} \otimes \left(U_q\mathfrak{b}^+\right)(F_N U_q\mathfrak{n}_-),$$

where $N = 0, 1, 2, \ldots$. Let $U_q\mathfrak{g}\hat{\otimes}U_q\mathfrak{g}$ be the completion of $U_q\mathfrak{g} \otimes U_q\mathfrak{g}$ with respect to this topology.

DEFINITION 7.2.2. (1) For any finite-dimensional $U_q\mathfrak{g}$-module V and $\lambda \in \mathfrak{h}^*$, define

$$V_\lambda = \{v \mid K_i v = q^{\langle \lambda, \alpha_i \rangle} v\}.$$

(2) A finite-dimensional $U_q\mathfrak{g}$-module V is said to be *admissible*, if it admits the decomposition

(7.2.1) $$V = \bigoplus_{\lambda \in \mathfrak{h}^*} V_\lambda.$$

(3) For any pair of admissible finite-dimensional $U_q\mathfrak{g}$-modules V and W, define $\Psi_{V,W} \in \operatorname{End}_{\mathbb{C}}(V \otimes W)$ by

(7.2.2) $$\Psi_{V,W} : v \otimes w \mapsto q^{(\lambda,\mu)} v \otimes w,$$

for any $v \otimes w \in V_\lambda \otimes W_\mu$.

THEOREM 7.2.3. *There exists an element* $\Theta \in U_q\mathfrak{g}\hat{\otimes}U_q\mathfrak{g}$ *with the following properties:*

(1) *There exists a decomposition*

$$\Theta = \sum_{\nu=0}^{\infty} \Theta_\nu,$$

where

$$\Theta_\nu \in (U_q\mathfrak{n}_+)_\nu (U_q\mathfrak{h}) \otimes (U_q\mathfrak{n}_-)_\nu (U_q\mathfrak{h})$$

and

$$\Theta_0 = 1 \otimes 1.$$

(2) *For any $a \in U_q\mathfrak{g}$ and a pair of finite-dimensional $U_q\mathfrak{g}$-modules V and W, we have*

(7.2.3) $$\Theta \Psi_{V,W} \Delta(a) = \Delta'(a) \Theta \Psi_{V,W}.$$

7.3. Admissible finite-dimensional $U_q\mathfrak{g}$-modules.

THEOREM 7.3.1. *The category $\mathcal{M}_q$ of admissible finite-dimensional $U_q\mathfrak{g}$-modules is a ribbon category.*

PROOF. The following is a sketch of the proof, whose details we leave to the reader as an exercise. It basically follows the course of the proof of Theorem 6.0.15, with only minor changes (like replacing $e^{\frac{h}{2}}$ by q and $e^{h\rho}$ by $\prod_{i=1}^{n} K_i^{m_i}$, where $\rho = \frac{1}{2}\sum_{i=1}^{n} m_i \alpha_i$) except for the following modification.

Since $U_q\mathfrak{g}$ is not a quasi-triangular Hopf algebra, an analog of the element u should be defined not by means of a genuine R-matrix, but by means of the element Θ defined in Theorem 7.2.3. Namely, let

$$\Theta = \sum_{i=0}^{\infty} a_i \otimes b_i,$$

where $a_i \in U_q\mathfrak{n}_+, b_i \in U_q\mathfrak{g}_-$. Define a sequence of elements u_k $(k = 0, 1, 2, \ldots)$ in $U_q\mathfrak{g}$ by

(7.3.1) $$u_k = \sum_{i=0}^{k} S(b_i) a_i.$$

For any admissible finite-dimensional $U_q\mathfrak{g}$-module V denote by $(u_k)_V$ the action of u_k in V. Also, define a linear map $\Gamma_V : V \to V$ by

$$\Gamma_V : v \mapsto q^{-(\lambda,\lambda)}v,$$

for any $v \in V$.

The following lemma holds.

LEMMA 7.3.2. (1) *For any admissible finite-dimensional $U_q\mathfrak{g}$-module V, the sequence $\{(u_k)_V\}$ defined by (7.3.1) stabilizes. That is, there exists an integer N such that $(u_m)_V = (u_n)_V$ for any $m, n \geq N$. In particular, this implies that the sequence converges to $u_V \in \text{End } V$.*

(2) *The linear map $\Gamma_V u_V : V \to V$ is a morphism of $U_q\mathfrak{g}$-modules. If $v \in V_\lambda$, then*

$$\Gamma_V u_V : v \mapsto q^{-(\lambda+2\rho,\lambda)}v.$$

It follows from Lemma 7.3.2 that $\Gamma_V u_V$ corresponds to the action of $q^{-2\rho}u$ in V. After that, one can complete the proof of Theorem 7.3.1 in the same way as the proof of Theorem 6.0.15. Namely, we find out that the braiding $S_{V,W} : V \otimes W \to W \otimes V$ can be defined by

$$S_{V,W} = P_{V,W} \Theta \Psi_{V,W},$$

where $P_{V,W} : v \otimes w \mapsto w \otimes v$ is the standard permutation. The balancing $b_V : V \to V$ can be defined by

$$b_V = \Gamma_V u_V.$$

The lemma is proved. $\square$

COROLLARY 7.3.3. *Define the matrix elements $c_{l,v}^\pi$ of a representation π in an admissible finite-dimensional $U_q\mathfrak{g}$-module V in the same way as it was done in Definition 5.1.7, replacing $U_h\mathfrak{g}$ by $U_q\mathfrak{g}$. Then the matrix elements of all admissible finite-dimensional representations of $U_q\mathfrak{g}$ form a Hopf subalgebra in the dual Hopf algebra*

$$(U_q\mathfrak{g})^* = \text{Hom}_{\mathbb{C}}(U_q\mathfrak{g}, \mathbb{C}).$$

7.4. Twisted version of $U_q\mathfrak{g}$. Let $v_0 \in \mathfrak{h} \wedge \mathfrak{h}$, $v = \exp h v_0$. Take an orthonormal basis $\{I_\mu\}$ in $\mathfrak{h}$. If
$$v_0 = \sum_{\mu,nu} v_0^{\mu\nu} I_\mu \otimes I_\nu,$$
then
$$v = q^{2\sum_{\mu,\nu} v_0^{\mu\nu} I_\mu \otimes I_\nu},$$
where $q = e^{\frac{h}{2}}$. Let us fix q as in Definition 7.1.1 and v_0 as above.

Define an automorphism $x \mapsto vxv^{-1}$ of the algebra $U_q\mathfrak{g} \otimes U_q\mathfrak{g}$ by

(7.4.1) $\quad v\left(X_i^\pm \otimes 1\right)v^{-1} = q^{\pm \sum_{\mu,\nu} v_0^{\mu\nu}(\alpha_i, I_\mu)}\left(X_i^\pm \otimes 1\right),$

(7.4.2) $\quad v\left(1 \otimes X_i^\pm\right)v^{-1} = q^{\pm \sum_{\mu,\nu} v_0^{\mu\nu}(\alpha_i, I_\nu)}\left(1 \otimes X_i^\pm\right),$

(7.4.3) $\quad v\left(K_i \otimes 1\right)v^{-1} = K_i \otimes 1,$

(7.4.4) $\quad v\left(1 \otimes K_i\right)v^{-1} = 1 \otimes K_i.$

Given the standard comultiplication $\Delta: U_q\mathfrak{g} \to U_q\mathfrak{g} \otimes U_q\mathfrak{g}$, we define the twisted comultiplication by
$$\Delta_{v_0}(\xi) = v\Delta(\xi)v^{-1},$$
for any $\xi \in U_q\mathfrak{g}$.

EXERCISE 7.4.1. Prove that Δ_{v_0} defines a Hopf algebra structure on $U_q\mathfrak{g}$.

DEFINITION 7.4.2. We denote this new Hopf algebra by $U_{q,v_0}\mathfrak{g}$ and call it the *twisted* (by v_0) quantized universal enveloping algebra.

Since $U_{q,v_0}\mathfrak{g}$ is isomorphic to $U_q\mathfrak{g}$ as an algebra, we can extend the notion of admissible module to the case of $U_{q,v_0}\mathfrak{g}$ (cf. Definition 7.2.2). Thus, we obtain a category $\mathcal{M}_{v_0}$ of finite-dimensional admissible $U_{q,v_0}\mathfrak{g}$-modules. Let
$$V = \bigoplus_{\lambda \in \mathfrak{h}^*} V_\lambda,$$
$$W = \bigoplus_{\mu \in \mathfrak{h}^*} W_\mu$$
be two finite-dimensional admissible $U_{q,v_0}\mathfrak{g}$-modules. Given $x_\lambda \in V_\lambda$ and $y_\mu \in W_\mu$, we define
$$v_{V,W}(x_\lambda \otimes y_\mu) = q^{\sum_{k,l} v_0^{kl}(\lambda, I_k)(\mu, I_l)} x_\lambda \otimes y_\mu,$$
$$v_{V,W}^{21}(x_\lambda \otimes y_\mu) = q^{\sum_{k,l} v_0^{kl}(\mu, I_k)(\lambda, I_l)} x_\lambda \otimes y_\mu$$
and extend these maps to linear endomorphisms $v_{V,W}$ and $v_{V,W}^{21}$ of $V \otimes W$. Finally, we define $\Theta_{v_0} \in \mathrm{End}_{\mathbb{C}}(V \otimes W)$ by
$$\Theta_{v_0}^{V,W} = v_{V,W}^{21} \Theta^{V,W} v_{V,W}^{-1},$$
where $\Theta^{V,W} \in \mathrm{End}_{\mathbb{C}}(V \otimes W)$ is the linear endomorphism defined by the quasi-R-matrix Θ from Section 7.2.

EXERCISE 7.4.3. Prove that the following equality holds in $\mathrm{End}_{\mathbb{C}}(V \otimes W)$:
$$\Theta_{v_0}^{V,W} \Psi_{V,W} \Delta_{v_0}(a) = \Delta'_{v_0} \Theta_{v_0}^{V,W} \Psi_{V,W},$$
where $\Psi_{V,W}$ was defined by (7.2.2).

EXERCISE 7.4.4. Prove that $\mathcal{M}_{v_0}$ is a rigid balanced category.

8. Historical remarks

The definition of the quantization of a Lie bialgebra is due to Drinfeld (cf. [**Dri87**]). In the case of a generalized symmetrizable Cartan matrix the formulas (1.2.1)–(1.2.6) were presented by Drinfeld in [**Dri85**] and independently by Jimbo in [**Jim86**], who considered the case of a fixed q.

Theorem 1.3.1 was conjectured by Drinfeld, and proved by Etingof and Kazhdan in [**EK96**].

Sections 2–5 reproduce the results of Drinfeld's paper [**Dri89**]. The twisted version of $U_h\mathfrak{g}$ (cf. Section 2.4) is due to Levendorskii and Soibelman (cf. [**LS91b**]). The formulas (5.1.7)–(5.1.9), which show that difference operators arise in the quantum groups theory, were discovered independently by a number of authors (see, for instance, [**Jim86, SV88**]).

For the standard background on monoidal and braided monoidal categories see the book [**Mac71**] by Mac Lane. As was mentioned earlier in this chapter, the material of Section 7 can be found in more detail and a slightly different notation in Lusztig's book [**Lus93**].

We would like to mention that nontwisted quantized universal enveloping algebras are discussed in every book on quantum groups. The case of fixed q (Section 7) is very popular because of applications to topology (knot invariants and invariants of three-dimensional manifolds) and to representations of Chevalley groups over finite fields. For the interested reader we can recommend the above-mentioned books [**Lus93**], [**Jant96**], or [**Kas95**]. An extensive list of references can be found in [**ChP95**].

CHAPTER 3

Quantized Algebras of Functions

Let G be a connected and simply connected simple complex Lie group, and $K \subset G$ its maximal compact subgroup defined in Chapter 1, Section 5.1. Given $q \in \mathbb{C} \setminus \{0\}$ which is not a root of unity, we will define a Hopf algebra $\mathbb{C}[G]_q$, the quantized algebra of regular functions, a quantum analog of $\mathbb{C}[G]$. If q is real, then one can define an involution on $\mathbb{C}[G]_q$ which makes it a Hopf $*$-algebra $\mathbb{C}[K]_q$, the quantized algebra of K-finite functions, a quantum analog of $\mathbb{C}[K]$ (the precise definitions will be given later).

In this chapter we study the structure of these algebras, as well as the representation theory of $\mathbb{C}[K]_q$ and its relation to the Bruhat decomposition (see Chapter 1, Section 5.2). All of the results admit versions over $\mathbb{C}[[h]]$, where we can define $\mathbb{C}[G]_h$ and $\mathbb{C}[K]_h$ in a similar way as we do it over the field of complex numbers. In the previous chapter we mostly studied $U_h \mathfrak{g}$, the quantized universal enveloping algebra over $\mathbb{C}[[h]]$, and only briefly mentioned the case $U_q \mathfrak{g}$ of the fixed q. Here we start with the case of fixed q and at the end mention the case of formal power series.

1. Main definitions

1.1. Hopf $*$-algebras. Let $\mathbf{k}$ be a commutative ring with unity, and A a $\mathbf{k}$-algebra with unity.

DEFINITION 1.1.1. Given an A-module V, a pair (l, v) with $l \in V^*$ and $v \in V$ defines a linear functional on A by $a \mapsto l(av)$. This linear functional is called the *matrix element* of the corresponding representation $\pi : A \to \text{End}_{\mathbf{k}} V$ corresponding to the pair (l, v). We will denote it by $c_{l,v}^{\pi}$ or sometimes by $c_{l,v}^{V}$.

REMARK 1.1.2. Throughout the rest of this section we will consider only those A-modules that are projective as $\mathbf{k}$-modules. We recall that if such a module V has a finite rank over $\mathbf{k}$, we will call V *finite-dimensional* A-module. If $\mathbf{k}$ is a topological algebra, then all the definitions should be modified to include the assumption of continuity.

PROPOSITION 1.1.3. *Let A be a Hopf algebra over $\mathbf{k}$. Then the matrix elements of finite-dimensional A-modules form a Hopf subalgebra A^* in the dual algebra $A^* = \text{Hom}_{\mathbf{k}}(A, \mathbf{k})$.*

PROOF. This is just a reformulation of Proposition 2.2.6. Generally speaking, this is a consequence of the fact that the finite-dimensional modules over a Hopf algebra form a rigid monoidal category in the sense of Chapter 2, Section 6. □

DEFINITION 1.1.4. We call A^* the *dual Hopf algebra* to A.

Suppose that $\mathbf{k}$ is equipped with an involutive algebra homomorphism $\lambda \mapsto \bar{\lambda}$, and $a \mapsto a^*$ is an involutive homomorphism of abelian groups $A \to A$.

DEFINITION 1.1.5. We say that $a \mapsto a^*$ is an *involution* of the **k**-algebra A if the following conditions are satisfied:
(1) $(\lambda a)^* = \bar{\lambda} a^*$ for any $\lambda \in \mathbf{k}$, $a \in A$;
(2) $(ab)^* = b^* a^*$ for any $a, b \in A$.

DEFINITION 1.1.6. Let A be a **k**-algebra with involution, and V an A-module. We say that V is *unitarizable* if it can be equipped with a Hermitian scalar product $\langle \cdot, \cdot \rangle : V \times V \to \mathbf{k}$ such that for any $a \in A, v_1, v_2 \in V$, we get

(1.1.1) $$\langle a v_1, v_2 \rangle = \langle v_1, a^* v_2 \rangle.$$

Also, we call $\langle \cdot, \cdot \rangle$ a *scalar product* on V (with values in **k**).

DEFINITION 1.1.7. Let A be a Hopf algebra over **k**. We say that a homomorphism of abelian groups $* : A \to A$, $a \mapsto a^*$ equips A with a structure of *Hopf $*$-algebra* if the following conditions are satisfied:
(1) A becomes a **k**-algebra with involution when equipped with $*$;
(2) $\Delta(a^*) = (\delta(a))^*$ for any $a \in A$, where $\Delta : A \to A \otimes A$ is the comultipication, and $(a \otimes b)^* = a^* \otimes b^*$ for any $a, b \in A$.

REMARK 1.1.8. One can check that the conditions of Definition 1.1.7 imply that $(S(S(a)^*))^* = a$ for any $a \in A$. In other words, the map $\omega : a \mapsto (S(a))^*$ is an involution. Here $S : A \to A$ is the antipode.

PROPOSITION 1.1.9. *Suppose that A is a Hopf $*$-algebra. Then the dual Hopf algebra A^* has a canonical Hopf $*$-algebra structure given by the involution $l \mapsto l^*$, where*

(1.1.2) $$l^*(a) = \overline{l(S(a)^*)}$$

for any $l \in A^*$ and $a \in A$.

PROOF. A straightforward computation. $\square$

PROPOSITION 1.1.10. *Suppose that A is a Hopf $*$-algebra. The matrix elements of finite-dimensional unitarizable A-modules form a Hopf $*$-subalgebra in A^*.*

PROOF. Suppose that V_1 and V_2 are unitarizable A-modules. Let $\langle \cdot, \cdot \rangle_1$ and $\langle \cdot, \cdot \rangle_2$ be the corresponding scalar products. Then $M_1 \otimes_{\mathbf{k}} M_2$ is also a unitarizable A-module, with a scalar product given by

$$\langle v_1 \otimes v_2, v_3 \otimes v_4 \rangle = \langle v_1, v_3 \rangle_1 \cdot \langle v_2, v_4 \rangle_2.$$

This immediately implies the statement of the proposition. $\square$

1.2. Quantized algebra of regular functions. Suppose that we are given a nonzero complex number $q \in \mathbb{C} \setminus \{0\}$ which is not a root of unity. Consider the standard Poisson Lie group structure on G. Recall that the quantization of the corresponding Lie bialgebra $\mathfrak{g}$ is the quantized universal enveloping algebra $U_q \mathfrak{g}$ introduced in the previous chapter.

DEFINITION 1.2.1. The Hopf algebra of matrix elements of finite-dimensional admissible $U_q \mathfrak{g}$-modules is called the *quantized algebra of regular functions* on the connected and simply connected Lie group G with the Lie algebra $\mathfrak{g}$. We will denote it by $\mathbb{C}[G]_q$.

REMARK 1.2.2. The quantized algebra of regular functions can be found in the literature under different names, such as the *quantum coordinate algebra* of G or the *algebra of regular functions on the quantum group* G. Sometimes, we too will use these synonims.

Suppose now that q is a nonzero real number. Equip the field $\mathbb{C}$ of complex numbers with the usual conjugation involution $z \mapsto \bar{z}$.

PROPOSITION 1.2.3. *If q is a nonzero real number, there exists a structure of Hopf $*$-algebra on the complex Hopf algebra $U_q\mathfrak{g}$. It is given by an involutive algebra antiautomorphism*

(1.2.1) $$\theta : a \mapsto S(\omega(a)),$$

where S is the antipode, and ω is the involutive automorphism of $U_q\mathfrak{g}$ uniquely defined on the generators $X_i^\pm, K_i$ ($i = 1, 2, \ldots, n$) by

(1.2.2) $$\omega\left(X_i^\pm\right) = X_i^\mp, \quad \omega(K_i) = K_i.$$

PROOF. Straightforward computation. □

By Proposition 1.1.9, the dual Hopf algebra carries a canonical Hopf $*$-algebra structure. We will denote by $*$ the involution on $\mathbb{C}[G]_q$ dual to ω.

DEFINITION 1.2.4. Let $K \subset G$ be the maximal compact subgroup of G, corresponding to the involution ω_0 defined in Chapter 1, Section 5.1. The *quantized algebra of functions* $\mathbb{C}[K]_q$ is the Hopf algebra $(\mathbb{C}[G]_q, *)$ dual to the Hopf $*$-algebra $(U_q\mathfrak{g}, *)$ considered in Proposition 1.2.3.

REMARK 1.2.5. (i) Similarly to Remark 1.2.2, $\mathbb{C}[K]_q$ is also called sometimes the *quantum coordinate algebra* of K or the *algebra of K-finite functions on the quantum group* K.

(ii) If we start with $\mathbf{k} = \mathbb{C}[[h]]$ instead of $\mathbf{k} = \mathbb{C}$, equipped with the involution $\sum_{n=0}^\infty a_n h^n \mapsto \sum_{n=0}^\infty \bar{a}_n h^n$, then we can repeat all the definitions and propositions above, with the additional requirement of continuity in the h-adic topology, and define the *formal quantized algebras of functions* $\mathbb{C}[G]_h$ and $\mathbb{C}[K]_h$.

2. Properties of the quantized algebras of functions

2.1. Basic properties. Let $V = L(\Lambda)$ be the simple admissible $U_q\mathfrak{g}$-module with the highest weight $\Lambda \in P_+$, where P_+ is the set of all dominant weights in $\mathfrak{h}^*$ (see Chapter 2, Section 7.3). Then we have the weight decomposition

(2.1.1) $$L(\Lambda) = \bigoplus_{\lambda \in \mathfrak{h}^*} L(\Lambda)_\lambda.$$

For the dual $U_q\mathfrak{g}$-module $L(\Lambda)^*$, we also have the weight decomposition

$$L(\Lambda)^* = \bigoplus_{\lambda \in \mathfrak{h}^*} L(\Lambda)^*_\lambda,$$

where

$$L(\Lambda)^*_\lambda = \left(L(\Lambda)_{-\lambda}\right)^*.$$

DEFINITION 2.1.1. Given any Hopf algebra A, the *right* (resp. *left*) *regular representation* is the action of A on A^* given by
$$\mathcal{R}(a) : f \mapsto \langle \mathrm{id} \otimes a, \Delta(f) \rangle,$$
$$\mathcal{L}(a) : f \mapsto \langle S^{-1}(a) \otimes \mathrm{id}, \Delta(f) \rangle,$$
where S is the antipode in A, and Δ is the comultiplication in A^*.

In particular, we can consider the left and right regular representations of $U_q\mathfrak{g}$ in $\mathbb{C}[G]_q$.

PROPOSITION 2.1.2. *Consider the action of $U_q\mathfrak{g} \otimes U_q\mathfrak{g}$ on $\mathbb{C}[G]_q$ given by $\mathcal{L} \otimes \mathcal{R}$. There exists a canonical isomorphism*

$$(2.1.2) \qquad \mathbb{C}[G]_q \simeq \bigoplus_{\Lambda \in P_+} L(\Lambda)^* \otimes L(\Lambda)$$

as $U_q\mathfrak{g} \otimes U_q\mathfrak{g}$-bimodules, given by

$$c_{l,v}^{L(\Lambda)} \mapsto l \otimes v.$$

PROOF. The assertion follows directly from the definition of $\mathbb{C}[G]_q$. □

In particular, it follows that there is a linear basis of $\mathbb{C}[G]_q$, consisting of the matrix elements $c_{\lambda\mu}^\Lambda$, where $c_{\lambda\mu}^\Lambda$ is a shortened notation for any matrix element of the type $c_{l,v}^{L(\Lambda)}$ such that $l \in L(\Lambda)^*_\lambda$ and $v \in L(\Lambda)_\mu$. Sometimes we will omit the upper index and write simply $c_{\lambda\mu}$ if it causes no ambiguity.

EXERCISE 2.1.3. Prove that $\mathbb{C}[G]_q$ is an integral domain.

PROPOSITION 2.1.4. *Let $\Lambda, \Lambda' \in P_+$ and $l \in L(\Lambda)^*_{-\lambda}$, $l' \in L(\Lambda')_{-\lambda'}$. Suppose that $v' \in L(\Lambda')_\gamma$ and that $v_\Lambda \in L(\Lambda)_\Lambda$ is a highest weight vector. Then there exist $a_\nu \in \mathbb{C}$, $l_\nu \in ((U_q\mathfrak{b}_+)l)_{-\lambda+\nu}$, $l'_\nu \in ((U_q\mathfrak{b}_-)l')_{-\lambda'-\nu}$ such that*

$$(2.1.3) \qquad c_{l',v'}^{\Lambda'} c_{l,v_\Lambda}^{\lambda} = q^{2((\Lambda,\gamma)-(\lambda,\lambda'))} c_{l,v_\Lambda}^{\Lambda} c_{l',v'}^{\Lambda'} + q^{2(\Lambda,\gamma)} \sum_{\nu \in Q_+} a_\nu c_{l_\nu,v_\Lambda}^{\Lambda} c_{l'_\nu,v'}^{\Lambda'},$$

where Q_+ is the lattice of positive roots of $\mathfrak{g}$.

PROOF. This proposition can be deduced from Theorem 7.2.3 of Chapter 2. □

COROLLARY 2.1.5. *The following identities hold:*

$$(2.1.4) \qquad c_{-\Lambda,w_0\lambda}^{\Lambda} c_{-w_0\Lambda',\Lambda'}^{\Lambda'} = c_{-w_0\Lambda',\Lambda'}^{\Lambda'} c_{-\Lambda,w_0\lambda}^{\Lambda},$$

$$(2.1.5) \qquad c_{-\mu,\lambda}^{\Lambda} c_{-w_0\Lambda',\Lambda'}^{\Lambda'} = q^{2((\lambda,\Lambda')-(\mu,w_0\Lambda'))} c_{-w_0\Lambda',\Lambda'}^{\Lambda'} c_{-\mu,\lambda}^{\Lambda},$$

$$(2.1.6) \qquad c_{-\Lambda,w_0\Lambda}^{\Lambda} c_{-\mu,\lambda}^{\Lambda'} = q^{2((\lambda,w_0\Lambda)-(\mu,\Lambda))} c_{-\mu,\lambda}^{\Lambda'} c_{-\Lambda,w_0\Lambda}^{\Lambda},$$

where w_0 is the element of the Weyl group W of maximal length, $\Lambda, \Lambda' \in P_+$, and λ, μ are the weights of the corresponding simple $U_q\mathfrak{g}$-modules.

PROOF. It is easy to see that if the weights of l, l', v' are specified as in (2.1.4)–(2.1.5), the extra sum on the right-hand side of (2.1.3) disappears, since there are no such weights in $L(\Lambda)^*$ and $L(\Lambda')^*$ as those which l_ν and l'_ν should have. This proves (2.1.4) and (2.1.5).

Another way to approach all three formulas is to start with (2.1.4) which can be checked by a straightforward computation. Applying $U_q\mathfrak{b}_+$ and $U_q\mathfrak{b}_-$ to both sides of (2.1.4), one can deduce (2.1.5) and (2.1.6).

Moreover, applying $U_q\mathfrak{b}_+$ and $U_q\mathfrak{b}_-$ to both sides of (2.1.6) and (2.1.5) respectively, one can give another proof of Proposition 2.1.4 by induction. □

2.2. Triangular decomposition of $\mathbb{C}[G]_q$. Let $A_+ \subset \mathbb{C}[G]_q$ be the subalgebra generated by the matrix elements of the form $c_{\lambda,\Lambda}^\Lambda$ ($\Lambda \in P_+$), and $A_- \subset \mathbb{C}[G]_q$ the subalgebra generated by the matrix elements of the form $c_{\lambda,w_0\Lambda}^{-w_0\Lambda}$ ($\Lambda \in P_+$). Imposing a stricter requirement on Λ to belong to the subset $P_{++} \subset P_+$ of regular dominant weights, we define the subalgebras $A_{++} \subset A_+$ and $A_{--} \subset A_-$.

Consider the involution $*$ which makes $\mathbb{C}[G]_q$ the Hopf algebra $\mathbb{C}[K]_q = (\mathbb{C}[G]_q, *)$ (see Proposition 1.2.3 and Definition 1.2.4). It is clear that the involution $*$ yields isomorphisms

$$A_+^* \simeq A_-, \quad A_{++}^* \simeq A_{--}.$$

We also have homomorphisms of complex vector spaces

$$m_+ : A_- \otimes A_+ \to \mathbb{C}[G]_q,$$
$$m_{++} : A_{--} \otimes A_{++} \to \mathbb{C}[G]_q$$

given by the multiplication $a \otimes b \mapsto ab$.

THEOREM 2.2.1. *Both m_+ and m_{++} are surjective.*

PROOF. We give a proof of the statement for m_+. The proof for m_{++} is similar. One just has to replace P_+ by P_{++} and A_+ by A_{++}.

First, consider the isomorphisms of $U_q\mathfrak{g}$-modules given by

$$\bigoplus_{\lambda \in P_+} L(\lambda)^* \xrightarrow{\sim} A_+, \quad l \mapsto l(\rho_\lambda(a)v_\lambda),$$

$$\bigoplus_{\mu \in P_+} L(-w_0\mu)^* \xrightarrow{\sim} A_+, \quad l \mapsto l(\rho_{-w_0\mu}(a)v_{-\mu}),$$

where $v_\lambda \in L(\lambda)$ is a (fixed) highest weight vector, while $v_{-\mu} \in L(-w_0\mu)$ is a (fixed) lowest weight vector. Thus, it suffices to show that m_+ is surjective as a linear map

$$m_+ : \bigoplus_{\lambda,\mu \in P_+} L(\lambda)^* \otimes L(-w_0\mu)^* \to \bigoplus_{\Lambda \in P_+} L(\Lambda)^* \otimes L(\Lambda).$$

Second, note that the product of two matrix elements is given by

$$l(\rho_\lambda(a)v_\lambda)l'(\rho_{-w_0\mu}(a)v_{-\mu}) = (l \otimes l')\left((\rho_\lambda \otimes \rho_{-w_0\mu})(a)v_\lambda \otimes v_{-\mu}\right).$$

We conclude that it suffices to prove that the linear map

$$\Phi : \bigoplus_{\lambda-\mu=\gamma, \lambda,\mu \in P_+} \mathrm{Hom}_{U_q\mathfrak{g}}\left(L(\lambda) \otimes L(-w_0\mu), L(\nu)\right) \to L(\nu)_\gamma$$

given by $\Phi(f) = f(v_\lambda \otimes v_{-\mu})$ is surjective for any $\nu \in P_+$ and $\gamma \in P(\nu)$.

For any $\lambda, \mu \in P_+$, consider the restriction $\Phi_{\lambda,\mu}$ of the map Φ to the subspace $\mathrm{Hom}_{U_q\mathfrak{g}}(L(\lambda) \otimes L(-w_0\mu), L(\nu))$. The following lemmas are well known (cf., for example, [**Lus93**], Propositions 31.2.6 and 3.5.6).

LEMMA 2.2.2. $\Phi_{\lambda,\mu}$ is an isomorphism on the subspace of $L(\nu)_\gamma$ of the vectors v such that $\left(X_i^+\right)^n v = 0$ for any $n > \lambda(H_i)$ and $\left(X_i^-\right)^n v = 0$ for any $n > \mu(H_i)$, $i = 1, 2, \ldots, n$.

LEMMA 2.2.3. If $\Lambda \in P_+$, then

$$L(\Lambda) \simeq U_q \mathfrak{n}_\pm \bigg/ \left(\sum_{i=1}^n U_q \mathfrak{n}_\pm\right) \left(X_i^\pm\right)^{\Lambda(H_i)+1},$$

where $U_q \mathfrak{n}_\pm \subset U_q \mathfrak{g}$ is the subalgebra generated by $X_i^\pm$, $i = 1, 2, \ldots, n$.

Therefore, if we take λ and μ such that $\lambda - \mu = \gamma$ and both $\lambda(H_i)$ and $\mu(H_i)$ are large enough for any $i = 1, 2, \ldots, n$, we get the whole weight space $L(\mu)_\gamma$ as the image of $\Phi_{\lambda,\mu}$. This concludes the proof of the theorem. □

The role of the above theorem in the representation theory of quantized algebras of function is similar to the role of the so-called triangular decomposition in the theory of $U\mathfrak{g}$-modules with highest weight vector, with $A_+ \subset \mathbb{C}[G]_q$ replacing a Borel subalgebra. Indeed, we show in Section 5 of this chapter how one can develop the representation theory for $\mathbb{C}[G]_q$ using the "highest weight" approach. As we will see there, there is no precise analog of the Cartan subalgebra for $\mathbb{C}[G]_q$. However, the following proposition shows that there is still a distinguished commutative subalgebra in $\mathbb{C}[G]_q$ which will play a similar role.

PROPOSITION 2.2.4. The elements $c^\Lambda_{-w_0\Lambda,\Lambda}$, $\Lambda \in P_+$, commute.

PROOF. Immediately follows from (2.1.5). □

2.3. The involution $*$ in $\mathbb{C}[K]_q$. Suppose that q is a nonzero real number and that $q \neq \pm 1$. Recall that in this case we have the involution $*$ on $\mathbb{C}[G]_q$ which defines the Hopf $*$-algebra $\mathbb{C}[K]_q$.

EXERCISE 2.3.1. Prove that $L(\Lambda)$ is a unitarizable $U_q\mathfrak{g}$-module for any $\Lambda \in P_+$ (see, for example, [**Lus93**, Chapter 6]).

Fix an orthonormal basis $\{v_\lambda^{(i)}\}$ in $L(\Lambda)$, where $\lambda \in P(\Lambda)$, $P(\Lambda)$ is the set of weights of $L(\Lambda)$, and $v_\lambda^{(i)} \in L(\Lambda)_\lambda$ ($i = 1, 2, \ldots, \dim L(\Lambda)_\lambda$). Consider the dual basis $\{l_{-\lambda}^{(i)}\}$ in $L(\Lambda)^*$. In the following proposition we denote by $c^\Lambda_{-\mu,i,\lambda,j}$ the matrix element of $L(\Lambda)$ corresponding to the pair $(l_{-\mu}^{(i)}, v_\lambda^{(j)})$.

PROPOSITION 2.3.2. One has
$$\left(c^\Lambda_{-\mu,i,\lambda,j}\right)^* = q^{2(\mu-\lambda,\rho)} c^{w_0\Lambda}_{\mu,i,-\lambda,j}.$$

PROOF. It is clear that the action of the involution $*$ on the matrix elements of the form $c^\Lambda_{-\mu,\lambda}$ does not depend on the choice of the orthonormal basis. Note that the basis can be chosen to be compatible with the isomorphism $L(\Lambda)^* \simeq L(-w_0\Lambda)$ of $U_q\mathfrak{g}$-modules. Then

(2.3.1) $$\left(c^\Lambda_{-\mu,i,\lambda,j}\right)^*(a) = \left(\pi_\Lambda(S(a))v_\mu^{(i)}, v_\lambda^{(j)}\right)_{L(\Lambda)}$$

for any $a \in U_q\mathfrak{g}$, where S is the antipode in $U_q\mathfrak{g}$, $(\ ,\ )_{L(\Lambda)}$ is the scalar product on $L(\Lambda)$ (which makes it a unitarizable $U_q\mathfrak{g}$-module), and π_Λ is the representation of $U_q\mathfrak{g}$ in $L(\Lambda)$.

Let $\theta : U_q\mathfrak{g} \to U_q\mathfrak{g}$ be the involutive algebra antiautomorphism given by (1.2.1). Recall that

(2.3.2) $$\theta\left(X_i^\pm\right) = -q_i^{\pm 1} X_i^\mp, \quad \theta(K_i) = K_i^{-1}.$$

Then $\pi_\Lambda \circ \theta$ defines a structure of a $U_q\mathfrak{g}$-module on the vector space $L(\Lambda)$. It is easy to see that it is isomorphic to $L(-w_0\Lambda)$ as $U_q\mathfrak{g}$-module. Now we define a new scalar product on this module which makes it a unitarizable one. Namely, for any vectors $v_\lambda^{(i)}$ and $v_\mu^{(j)}$ of the basis, we define the scalar product by

(2.3.3) $$(v_\lambda^{(i)}, v_\mu^{(j)})_{\text{new}} = q^{(\mu,\rho)} (v_\lambda^{(i)}, v_\mu^{(j)})_{\text{old}}.$$

It is easy to check that this gives us a unitarizable $U_q\mathfrak{g}$-module. By construction, the vectors

(2.3.4) $$\check{v}_{-\lambda}^{(i)} = q^{2(\lambda,\rho)} v_\lambda^{(i)}$$

form an orthonormal basis compatible with the weight decomposition. Finally, the statement of the proposition follows from (2.3.4) and (2.3.1). $\square$

EXAMPLE 2.3.3. For the quantum universal enveloping algebra $U_q\mathfrak{sl}(2,\mathbb{C})$, the following formulas define the $m+1$-dimensional standard representation ρ_m of $U_q\mathfrak{sl}(2,\mathbb{C})$:

(2.3.5) $$\rho_m(X) : e_k \mapsto \frac{q^{k-m} - q^{-k+m}}{q - q^{-1}} e_{k+1},$$

(2.3.6) $$\rho_m(Y) : e_k \mapsto \frac{q^k - q^{-k}}{q - q^{-1}} e_{k-1},$$

(2.3.7) $$\rho_m(H) : e_k \mapsto (2k - m) e_k,$$

where $k = 0, 1, \ldots, m$ (cf. Chapter 2, (5.1.7)–(5.1.9)). Note that the representations ρ_m form a complete list of all irreducible admissible representations of $U_q\mathfrak{sl}(2,\mathbb{C})$ up to equivalence.

Consider the matrix $\begin{bmatrix} t_{11} & t_{12} \\ t_{21} & t_{22} \end{bmatrix}$ formed by the matrix elements of the standard two-dimensional representation. Then we have that

(2.3.8) $$t_{11}^* = t_{22}, \quad t_{21}^* = -q t_{12}.$$

3. Examples: $\mathbb{C}[SL_2(\mathbb{C})]_q$ and $\mathbb{C}[SU(2)]_q$

THEOREM 3.0.1. (1) *The quantized algebra of functions $\mathbb{C}[SL_2(\mathbb{C})]_q$ is generated by t_{ij}, $i,j = 1, 2$, subject to the following relations:*

(3.0.1) $\quad t_{11}t_{12} = q^{-1} t_{12}t_{11}, \qquad t_{11}t_{21} = q^{-1} t_{21}t_{11},$

(3.0.2) $\quad t_{12}t_{22} = q^{-1} t_{22}t_{12}, \qquad t_{21}t_{22} = q^{-1} t_{22}t_{21},$

(3.0.3) $\quad t_{12}t_{21} = t_{21}t_{12}, \qquad t_{11}t_{22} - t_{22}t_{11} = \left(q^{-1} - q\right) t_{12}t_{21},$

(3.0.4) $\qquad\qquad\qquad\qquad\qquad t_{11}t_{22} - q^{-1} t_{12}t_{21} = 1.$

(2) *The comultiplication Δ, the counit ε, and the antipode S in $\mathbb{C}[SL_2(\mathbb{C})]_q$ are given by*

(3.0.5) $$\Delta(t_{ij}) = \sum_{k=1,2} t_{ik} \otimes t_{kj}, \quad \varepsilon(t_{ij}) = \delta_{ij},$$

(3.0.6) $$S(t_{11}) = t_{22},\ S(t_{22}) = t_{11},\ S(t_{12}) = -qt_{12},\ S(t_{21}) = -q^{-1}t_{21}.$$

PROOF. The second part of the theorem easily follows from the definitions. We will prove the first part here. The proof that we give consists of three steps.

Step 1. Let t_{ij} $(i,j = 1,2)$ be the matrix elements of the standard two-dimensional representation π_1 of $U_q\mathfrak{sl}(2,\mathbb{C})$ described in Example 2.3.3. We claim that they satisfy relations (3.0.1)–(3.0.4). To prove this, let us consider the following matrix:

(3.0.7) $$\check{R} = \begin{pmatrix} q^{-2} & 0 & 0 & 0 \\ 0 & 0 & q^{-1} & 0 \\ 0 & q^{-1} & q^{-2}-1 & 0 \\ 0 & 0 & 0 & q^{-2} \end{pmatrix}.$$

View $\check{R}$ as an element of $\mathrm{End}\,(\mathbb{C}^2 \otimes \mathbb{C}^2)$. It is easy to check that it has the following properties:

(3.0.8) $$\check{R}^{12}\check{R}^{23}\check{R}^{12} = \check{R}^{23}\check{R}^{12}\check{R}^{23},$$

(3.0.9) $$\left(\check{R} - q^{-2}\right)\left(\check{R} + 1\right) = 0,$$

(3.0.10) $$\check{R}\,(\pi_1 \otimes \pi_1)(a) = (\pi_1 \otimes \pi_1)(a)\check{R},$$

for any $a \in U_q\mathfrak{sl}(2,\mathbb{C})$ (see Chapter 1, Section 4.1 for the notation $\check{R}^{ij}$).

The property (3.0.10) can be rewritten as

(3.0.11) $$\check{R}(T \otimes 1)(1 \otimes T) = (1 \otimes T)(T \otimes 1)\check{R},$$

where T is the 2×2 matrix with the entries t_{ij}. It is easy to see that (3.0.1)–(3.0.3) follow directly from (3.0.11).

To prove (3.0.4), consider the one-dimensional subrepresentation in $\pi_1 \otimes \pi_1$— the one generated by the so-called *q-skew-symmetrical tensors*, that is, $\mathrm{Ker}\,(\check{R}+1)$. This subrepresentation is equivalent to the trivial representation of $U_q\mathfrak{sl}(2,\mathbb{C})$—the one given by the counit

$$\varepsilon\left(X^\pm\right) = 0, \quad \varepsilon(K) = \varepsilon(1) = 1.$$

Restricting $\pi_1 \otimes \pi_1$ to $\mathrm{Ker}\,(\check{R}+1)$, we get (3.0.4). By the above construction, we see that the relation (3.0.4) can be thought of as a quantum analog of the determinant of the 2×2 matrix $T = (t_{ij})$ being equal 1.

Step 2. Let B be the free complex algebra with unity generated by x_{ij} $(i,j = 1,2)$. Let $\iota : B \to \mathbb{C}[SL_2(\mathbb{C})]_q$ be the natural algebra homomorphism such that $\iota(x_{ij}) = t_{ij}$. Indeed, what we have done in the previous step is precisely to show that this homomorphism does exist (of course, it is unique). Note that B can be equipped with a Hopf algebra structure, with the comultiplication and the counit given by

$$\bar{\Delta}(x_{ij}) = \sum_{k=1,2} x_{ik} \otimes x_{kj}, \quad \bar{\varepsilon}(x_{ij}) = \delta_{ij}.$$

The antipode $\bar{S}$ is uniquely determined by $\bar{\Delta}$ and $\bar{\varepsilon}$ and is given by (3.0.6) with t_{ij} replaced by x_{ij}. Therefore, ι is also a Hopf algebra homomorphism.

Any finite-dimensional representation of $U_q\mathfrak{sl}(2,\mathbb{C})$ is completely reducible (thus, it can be decomposed into a sum of irreducible ones), and any irreducible admissible representation π_l occurs in some tensor power of π_1. Therefore, ι is surjective. Also, it was shown in the previous step that $\iota(J) = 0$, where $J \subset B$ is a two-sided ideal generated by the relations (3.0.1)–(3.0.4), with t_{ij} replaced by x_{ij}.

The algebras B and $\mathbb{C}[SL_2(\mathbb{C})]_q$ admit filtrations

$$0 \subset B_0 \subset B_1 \subset \cdots \subset B,$$
$$0 \subset A_0 \subset A_1 \subset \cdots \subset A = \mathbb{C}[SL_2(\mathbb{C})]_q,$$

where

- the subspace B_m consists of noncommutative polynomials in x_{ij} of degree at most m;
- the subspace A_m is spanned by the matrix elements of the admissible representations of $U_q\mathfrak{sl}(2,\mathbb{C})$ of dimensions at most $m+1$.

It is clear that $\iota(B_N) \subset A_N$ for any N. We want to prove that ι induces an isomorphism $B/J \simeq \mathbb{C}[SL_2(\mathbb{C})]_q$. It suffices to show that for any N, the induced linear map

$$\iota_N : B_N / (B_N \cap (B_{N-1} + J)) \to A_N / A_{N-1}$$

is an isomorphism. It is easy to see that ι_N is surjective. We will show that

(3.0.12) $\quad \dim_\mathbb{C} B_N / (B_N \cap (B_{N-1} + J)) \leq (N+1)^2,$

(3.0.13) $\quad \dim_\mathbb{C} A_N / A_{N-1} \geq (N+1)^2,$

which completes the proof of the theorem.

Step 3 consists in the proof of (3.0.12)–(3.0.13). The following lemma is a direct consequence of (3.0.1)–(3.0.4).

LEMMA 3.0.2. *The vector space $B_N / (B_N \cap (B_{N-1} + J))$ is generated by the monomials belonging to any of the following three sets:*

$$Q_0 = \{x_{12}^{k_1} x_{21}^{k_2} \,|\, k_1 + k_2 = N\},$$
$$Q_1 = \{x_{11}^{k_1} x_{12}^{k_2} x_{21}^{k_3} \,|\, k_1 > 0, k_1 + k_2 + k_3 = N\},$$
$$Q_2 = \{x_{22}^{k_1} x_{12}^{k_2} x_{21}^{k_3} \,|\, k_1 + k_2 + k_3 = N\}.$$

Counting the elements in the sets Q_0, Q_1, and Q_3, we obtain (3.0.12). To prove (3.0.13), one can take a Casimir element $C_q \in U_q\mathfrak{sl}(2,\mathbb{C})$ given by

$$C_q = \frac{1}{2}\left(X^+X^- + X^-X^+\right) + \frac{q + q^{-1}}{2}\left(\frac{K - K^{-1}}{q - q^{-1}}\right)^2$$

(cf. Chapter 2, (4.1.3)) and check that it separates the irreducible admissible representations of $U_q\mathfrak{sl}(2,\mathbb{C})$ of different dimensions. This concludes the proof of the theorem. $\square$

COROLLARY 3.0.3. *Suppose that q is a nonzero real number. Then $\mathbb{C}[SU(2)]_q$ is isomorphic to the complex $*$-algebra generated by t_{11}, t_{21}, subject to the following relations:*

(3.0.14) $\quad t_{11}t_{21} = q^{-1}t_{21}t_{11}, \quad t_{11}t_{21}^* = q^{-1}t_{21}^*t_{11},$

(3.0.15) $\quad t_{21}t_{21}^* = t_{21}^*t_{21}, \quad t_{21}t_{11}^* = q^{-1}t_{11}^*t_{21},$

(3.0.16)

$$t_{11}t_{11}^* - t_{11}^*t_{11} = (q^2 - 1)t_{21}t_{21}^*, \quad t_{11}^*t_{11} + t_{21}^*t_{21} = 1.$$

PROOF. The assertion follows from Theorem 3.0.1 and (2.3.8). □

EXERCISE 3.0.4. Suppose that $R \in \mathrm{End}\,(\mathbb{C}^n \otimes \mathbb{C}^n)$ is given by

$$R = q\sum_{i=1}^n E_{ii} \otimes E_{ii} + \sum_{i\neq j} E_{ii} \otimes E_{jj} + (q - q^{-1})\sum_{i>j} E_{ij} \otimes E_{ji},$$

where E_{ij} is an $n \times n$ matrix whose only nonzero entry equals 1 and is located in the ith row and jth column. Consider

$$\check{R} = PR,$$

where $P : \mathbb{C}^n \otimes \mathbb{C}^n \to \mathbb{C}^n \otimes \mathbb{C}^n$ is the permutation map.

(1) Prove that $\mathbb{C}[SL_n(\mathbb{C})]_q$ is isomorphic to the algebra with unity generated by t_{ij} $(i, j = 1, 2, \ldots, n)$, subject to the relations

(3.0.17) $$\check{R}(T \otimes 1)(1 \otimes T) = (1 \otimes T)(T \otimes 1)\check{R},$$
(3.0.18) $$\mathrm{det}_q T = 1,$$

where $T = ((t_{ij}))_{i,j=1}^n$ is the matrix with the entries t_{ij},

$$T \otimes 1 = T \otimes I_n, \quad 1 \otimes T = I_n \otimes T,$$

I_n is the $n \times n$ identity matrix, and

$$\mathrm{det}_q T = \sum_{\sigma \in S_n}(-q)^{l(\sigma)} t_{1,\sigma(1)} t_{2,\sigma(2)} \cdots t_{n,\sigma(n)},$$

where S_n is the permutation group in n elements, and $l(\sigma)$ is the length of a permutation σ.

EXERCISE 3.0.5. Suppose that q is a nonzero real number. Prove that the involution $*$ in $\mathbb{C}[SU(n)]_q = (\mathbb{C}[SL_n(\mathbb{C})]_q, *)$ is given by

(3.0.19) $$t_{ij}^* = (-q)^{i-j} M_{ij},$$

where M_{ij} is the q-minor

$$M_{ij} = \mathrm{det}_q((t_{ab}))_{a\neq i, b\neq j}.$$

Prove that (3.0.18) can be rewritten as

(3.0.20) $$\sum_{i=1}^n t_{i1}^* t_{i1} = 1.$$

4. Representation theory of $\mathbb{C}[SU(2)]_q$

While the subsequent section will be devoted to a more general case of the representation theory of $\mathbb{C}[K]_q$, where K is a connected and simply connected simple compact Lie group, it will be based largely on the results of the present section where the basic case of $\mathbb{C}[SU(2)]_q$ is considered. On the other hand, the ideas used in this section will be generalized, so that it can be also considered as an illustration to the general theory.

Throughout this section we assume that q is a fixed real number, $0 < q < 1$.

PROPOSITION 4.0.1. *Let V be a unitarizable $\mathbb{C}[SL_2(\mathbb{C})]_q$-module. Then for any $v \in V$, we have*

(4.0.1) $$\langle t_{11}(v), t_{11}(v)\rangle \leq \langle v, v\rangle,$$
(4.0.2) $$\langle t_{21}(v), t_{21}(v)\rangle \leq \langle v, v\rangle,$$

where $\langle \cdot, \cdot \rangle$ is the scalar product in V.

PROOF. By (3.0.16), we have

(4.0.3) $$\langle t_{11}^* t_{11} v, v\rangle + \langle t_{21}^* t_{21} v, v\rangle = \langle v, v\rangle.$$

Applying (1.1.1) to (4.0.3), we get (4.0.1)–(4.0.2). □

Consider a unitarizable $\mathbb{C}[SL_2(\mathbb{C})]_q$-module V. Its completion is a separable Hilbert space H_V. Proposition 4.0.1 shows that the action of t_{11} and t_{21} can be extended to bounded linear operators $\pi_V(t_{11})$ and $\pi_V(t_{21})$ in H_V, thus giving rise to a representation $\pi_V : \mathbb{C}[SU(2)]_q \to \operatorname{End} H_V$ (that is, a $*$-representation of $\mathbb{C}[SL_2(\mathbb{C})]_q$), such that any $f \in \mathbb{C}[SU(2)]_q$ is represented by a bounded linear operator $\pi_V(f) : H_V \to H_V$. We will say that π_V is an irreducible representation if it is topologically irreducible (that is, contains no proper closed invariant subspaces).

REMARK 4.0.2. Recall that in Section 2.1 of this chapter we have introduced the subalgebras A_+ and A_{++} of $\mathbb{C}[G]_q$ which will play an important role in the general representation theory of $\mathbb{C}[K]_q$. Note that in the case $K = SU(2)$, both A_+ and A_{++} coincide with the subalgebra of $\mathbb{C}[SU(2)]_q$ generated by t_{11} and t_{21}.

4.1. Unitarizable simple $\mathbb{C}[SL_2(\mathbb{C})]_q$-modules.

PROPOSITION 4.1.1. *Suppose that $t \in \mathbb{C}, |t| = 1$. Then the following holds.*

(1) *There is a one-dimensional unitarizable $\mathbb{C}[SL_2(\mathbb{C})]_q$-module, with the action of the generators given by*

(4.1.1) $$\tau_t(t_{11}) = t, \qquad \tau_t(t_{22}) = t^{-1},$$
(4.1.2) $$\tau_t(t_{21}) = 0, \qquad \tau_t(t_{12}) = 0.$$

(2) *There is an infinite-dimensional unitarizable simple $\mathbb{C}[SL_2(\mathbb{C})]_q$-module with an orthonormal basis $\{e_k\}_{k=0}^\infty$, the action of the generators given by*

(4.1.3) $$\pi_t(t_{11}) : e_0 \mapsto 0,$$
(4.1.4) $$\pi_t(t_{11}) : e_k \mapsto \sqrt{1-q^2} e_{k-1}, \quad k > 0,$$
(4.1.5) $$\pi_t(t_{22}) : e_k \mapsto \sqrt{1-q^2} e_{k+1}, \quad k \geq 0,$$
(4.1.6) $$\pi_t(t_{21}) : e_k \mapsto t q^k e_k, \quad k \geq 0,$$
(4.1.7) $$\pi_t(t_{12}) : e_k \mapsto -t^{-1} q^{k+1} e_k, \quad k \geq 0.$$

PROOF. The first statement is obvious. To prove (2), note first that (4.1.3)–(4.1.7) define the structure of a $\mathbb{C}[SL_2(\mathbb{C})]_q$-module on the span V of the vectors e_k ($k = 0, 1, 2, \ldots$). Indeed, it suffices to check that all the relations (3.0.1)–(3.0.4) are satisfied. It is also easy to see that the scalar product in H makes V a unitarizable $\mathbb{C}[SL_2(\mathbb{C})]_q$-module (it is a straightforward computation to check that (1.1.1) holds).

Now we have to prove that V is a simple module. Suppose that $W \subset V$ is a nonzero $\mathbb{C}[SL_2(\mathbb{C})]_q$-submodule.

CLAIM 4.1.2. W contains e_m for some m.

To prove the claim, consider a nonnegative integer l and a vector $v \in W$ of the form
$$v = \sum_{j=1}^{n} c_{p_j} e_{p_j},$$
where all the coefficients c_{p_j} are nonzero complex numbers such that all the numbers $c_{p_j} q^{lp_j}$, $j = 1, 2, \ldots, n$, are distinct. Of course, such a vector v always exists. Then for $k = 0, 1, \ldots, n-1$ we have
$$\pi_t\left(t_{21}^{kl}\right)(v) = \sum_{j=1}^{n} c_{p_j} q^{klp_j} t^{kl} e_{p_j}.$$
Thus, we get a system of n equations of the form
$$u_k = \sum_{j=1}^{n} a_{jk} e_{p_j} \ (k = 0, 1, \ldots, n-1),$$
where $u_k \in W$ for any k and $\det((a_{jk})) \neq 0$. Therefore, all the vectors e_{p_j}, $j = 1, 2, \ldots, n$, belong to W. Take one of them for e_m. The claim is proved.

Now, if $e_m \in W$, then, by (4.1.4),
$$e_0 = \left(1 - q^2\right)^{-\frac{k}{2}} \pi_t\left(t_{11}^{m-1}\right) e_{m-1} \in W.$$
Hence, $e_0 \in W$. But e_0 generates V as a $\mathbb{C}[SL_2(\mathbb{C})]_q$-module. Therefore, $W = V$. □

REMARK 4.1.3. Formulas (4.1.1)–(4.1.7) define simple $\mathbb{C}[SL_2(\mathbb{C})]_q$-modules for any q which is not a root of unity. However, this statement remains valid and the proof remains the same if we allow q to take any complex value except $0, 1, -1$. Of course, (4.1.3)–(4.1.7) give a unitarizable simple module only if q is real.

PROPOSITION 4.1.4. Let V be a unitarizable simple $\mathbb{C}[SL_2(\mathbb{C})]_q$-module such that t_{21} has an eigenvector in V. If the corresponding eigenvalue is nonzero, there exists an orthonormal basis $\{e_k\}_{k=0}^{\infty}$ in V such that the action of $\mathbb{C}[SL_2(\mathbb{C})]_q$ in V is given by (4.1.3)–(4.1.7) for some $t \in \mathbb{C}$ such that $|t| = 1$. If the eigenvalue is zero, then V is one-dimensional.

PROOF. Let us fix a structure of a unitarizable $\mathbb{C}[SL_2(\mathbb{C})]_q$-module on V. Denote the corresponding scalar product by $\langle \cdot, \cdot \rangle$. Suppose that there exists $v_0 \in V$ such that $t_{21}(v_0) = \lambda_0 v_0$ with $\lambda_0 \neq 0$. Then

(4.1.8) $\qquad t_{21} t_{11}^k(v_0) = \lambda_0 q^{-k} t_{11}^k(v_0)$

for any $k = 0, 1, \ldots$. By (4.0.2),

(4.1.9) $\qquad |\lambda_0|^2 q^{-2k} \langle t_{11}^k(v_0), t_{11}^k(v_0)\rangle \leq \langle t_{11}^k(v_0), t_{11}^k(v_0)\rangle.$

Since $0 < q < 1$, there exists a nonnegative integer n such that $|\lambda_0|^2 q^{-2n} > 1$. By (4.1.9) with $k = n$, we get that $t_{11}^n(v_0) = 0$. Let m be the largest nonnegative integer such that $t_{11}^m(v_0) \neq 0$. By normalizing the vector $t_{11}^m(v_0)$, we get that there exists a vector $e_0 \in V$ such that $\langle e_0, e_0 \rangle = 1$ and $t_{11}(e_0) = 0$. It follows from (4.1.8) that e_0 is an eigenvector of t_{21} with a nonzero eigenvalue μ_0. Also, from (3.0.16) and (2.3.8) it follows that $|\mu_0| = 1$ and $t_{12}(e_0) = -\mu_0^{-1} q e_0$.

As e_0 is a common eigenvector for t_{21} and t_{12}, it follows from (3.0.1)–(3.0.4) that the linear span W of the vectors $t_{22}^k(e_0)$ ($k = 0, 1, \ldots$) is a unitarizable $\mathbb{C}[SL_2(\mathbb{C})]_q$-submodule of V. Since V is simple, $V = W$. We leave it as a simple exercise to check that $t_{22}^k(e_0) \neq 0$ for any $k \geq 0$. Hence, all the vectors $t_{22}^k(e_0)$ are linearly independent. Let $\mu_0 = t$. By normalizing the basis $\{t_{22}^k(e_0)\}_{k=0}^\infty$ of V, we obtain (4.1.3)–(4.1.7).

Finally, if $\lambda_0 = 0$, then $\operatorname{Ker} t_{21}$ is an invariant submodule in V. Therefore, V is one-dimensional. $\square$

PROPOSITION 4.1.5. *A unitarizable one-dimensional* $\mathbb{C}[SL_2(\mathbb{C})]_q$-*module* V *is isomorphic to the module given by* (4.1.1)–(4.1.2) *for some* $t \in \mathbb{C}$ *such that* $|t| = 1$.

PROOF. It follows from (3.0.16) that either t_{21} or t_{11} acts trivially in V, while the other operator is the multiplication by a complex number z such that $|z| = 1$. However, if $t_{11} = 0$ on V, then we see from (3.0.3) and (2.3.8) that $t_{21} = 0$ on V as well. Therefore, t_{11} cannot act trivially in V, so that $t_{21} = 0$. The rest of the proof is obvious. $\square$

4.2. Irreducible $*$-representations of $\mathbb{C}[SU(2)]_q$. Suppose that V is a unitarizable $\mathbb{C}[SL_2(\mathbb{C})]_q$-module. The completion $H_V = \bar{V}$ is a separable Hilbert space. Obviously, it yields a representation $\pi_V : \mathbb{C}[SU(2)]_q \to \operatorname{End} H_V$ (see Proposition 4.0.1). In particular, the two unitarizable simple $\mathbb{C}[SL_2(\mathbb{C})]_q$-modules considered in Proposition 4.1.1 yield representations of $\mathbb{C}[SU(2)]_q$ which will be denoted by τ_t and π_t respectively.

THEOREM 4.2.1. *Let* $\pi : \mathbb{C}[SU(2)]_q \to \operatorname{End} H$ *be an irreducible representation in a separable Hilbert space* H. *Then the following statements are true:*

(1) *There exists a unique straight line* $l \subset H$ *invariant with respect to* $\pi(A_+)$.

(2) *The representation* π *is unitary equivalent to one of the representations* τ_t *or* π_t *for some* $t \in \mathbb{C}$ *such that* $|t| = 1$.

PROOF. If $\operatorname{Ker} \pi(t_{21}) \neq 0$, then this is an invariant subspace in H. Since π is irreducible, this implies that $\pi(t_{21}) = 0$. It is easy to see that in this case π must be one-dimensional and, thus, unitary equivalent to one of the representations τ_t.

Suppose that $\operatorname{Ker} \pi(t_{21}) = 0$. By (2.3.8) and (3.0.3),

$$\pi(t_{21})\pi\left(t_{21}^*\right) = \pi\left(t_{21}^*\right)\pi(t_{21}), \tag{4.2.1}$$

that is, $\pi(t_{21})$ is a normal bounded linear operator in H. Hence, we can apply the general spectral theorem to $\pi(t_{21})$ (see, for example, [**Rud73**]). Denote by $E(\Omega)$ the spectral projection of $\pi(t_{21})$ which corresponds to a Borel subset Ω in $\mathbb{C}$. It follows from (3.0.1) that for any Ω, we have

$$\pi(t_{11})E(\Omega) = E(q\Omega)\pi(t_{11}). \tag{4.2.2}$$

Similarly, we get that

$$\pi\left(t_{11}^*\right)E(\Omega) = E\left(q^{-1}\Omega\right)\pi\left(t_{11}^*\right). \tag{4.2.3}$$

Let λ_1 and λ_2 be nonzero complex numbers belonging to the spectrum of $\pi(t_{21})$.

CLAIM 4.2.2. *There exists an integer* m *such that* $\lambda_2 = q^m \lambda_1$.

Indeed, if this is not the case, then one can find an open set $\Omega \subset \mathbb{C}$ such that $\lambda_1 \in \Omega$, $\lambda_2 \notin \Omega$, $q\Omega = \Omega$. By (4.2.2)–(4.2.3) we have $\pi(f)E(\Omega) = E(\Omega)\pi(f)$ for any $f \in \mathbb{C}[SU(2)]_q$. Thus, $E(\Omega)$ is a proper closed invariant subspace in H, which contradicts the assumption that π is irreducible. The claim is proved.

The claim shows that the spectrum of $\pi(t_{21})$ is contained in the set $\{\lambda q^k\}_{k=0}^{\infty} \cup \{0\}$ for some $z \in \mathbb{C}$. Then any straight line $l \subset \text{Ker}(\pi(t_{21}) - \lambda)$ satisfies the first statement of the theorem.

Assume that there are two such straight lines l_1 and l_2 corresponding to some eigenvalues z_1 and z_2 respectively. We leave it as an easy exercise to show that $z_1 = z_2$. Consider $v_i \in \text{Ker}\,\pi(t_{11}) \cap \text{Ker}(\pi(t_{11}) - z_i)$ $(i = 1, 2)$ such that $\langle v_1, v_2 \rangle = 0$ and $\langle v_1, v_1 \rangle = \langle v_2, v_2 \rangle = 1$. Then v_1 and v_2 generate closed subspaces H_1 and H_2 respectively in H which are both invariant with respect to $\pi\,(\mathbb{C}[SU(2)]_q)$. This contradicts the assumption that π is irreducible. Therefore, the line l is unique, and the statement (1) of the theorem is proved.

In order to prove (2), consider a vector $e_0 \in l$ such that $\langle e_0, e_0 \rangle = 1$. Applying (3.0.1)–(3.0.4), we obtain (4.1.3)–(4.1.7). We leave it as a simple exercise to check that $\pi(t_{21})e_0 = \mu e_0$ where $|\mu| = 1$. Then we see that π is unitary equivalent to π_μ. $\square$

DEFINITION 4.2.3. Let π be an irreducible representation of $\mathbb{C}[SU(2)]_q$. The homomorphism $\chi : A_+ \to \mathbb{C}$ defined by

$$\pi(f)v = \chi(f)v$$

for any $v \subset l$, where l is the invariant line of Theorem 4.2.1, is called the *highest weight* of π.

REMARK 4.2.4. It follows from Theorem 4.2.1 that any irreducible representation π of $\mathbb{C}[SU(2)]_q$ in a separable Hilbert space has the highest weight and is uniquely determined by this highest weight. It also follows that the space of π is the completion of a unitarizable simple $\mathbb{C}[SL_2(\mathbb{C})]_q$-module of the form described in Proposition 4.1.1.

Therefore, by Proposition 4.1.4, we have a one-to-one correspondence between the set of equivalence classes of irreducible representations of $\mathbb{C}[SU(2)]_q$ in a separable Hilbert space and the set of equivalence classes of unitarizable simple $\mathbb{C}[SL_2(\mathbb{C})]_q$-modules which have an eigenvector for t_{21}.

REMARK 4.2.5. There is an obvious one-to-one correspodence between the irreducible representations of $\mathbb{C}[SU(2)]_q$ described in Theorem 4.2.1 and the symplectic leaves of the Poisson Lie group $SU(2)$ described in Chapter 1, Example 5.2.4. Namely, the one-dimensional representations τ_t are parameterized by the zero-dimensional symplectic leaves—the points of the distinguished maximal torus,—while the infinite-dimensional representations π_t correspond to the two-dimensional symplectic leaves S_t (see (5.2.4)). Basically, the dimension of a representation corresponds to the "number of points" in the corresponding symplectic leaf.

We will return to this subject in the next section, where we will discuss it in more detail in the general context of a connected and simply connected simple compact Lie group K (equipped with the standard Poisson Lie group structure). Note, however, that the intuitive observation that $\frac{t_{21}}{t_{12}}$ "belongs" to the center of $\mathbb{C}[SL_2(\mathbb{C})]_q$ (or rather, to a certain localization of the center) was the starting point for the whole subject (cf. [**SV88**]).

5. Representation theory of $\mathbb{C}[K]_q$

Recall that we defined the subalgebras A_+ and A_{++} of $\mathbb{C}[G]_q$ in Section 2.2. Given a homomorphism $\chi : A_+ \to \mathbb{C}$ of algebras with unity, denote by $\mathbb{C}_\chi$ the corresponding A_+-module, and by $\mathbf{1}_\chi$ a generator of A_+. Below we develop the representation theory of $\mathbb{C}[K]_q$ in terms of highest weight modules, with A_+ playing the role of a Borel subalgebra. We assume that $0 < q < 1$.

5.1. Highest weight modules and primitive ideals.

DEFINITION 5.1.1. (1) We say that a $\mathbb{C}[G]_q$-module V is a *highest weight module* with *highest weight* $\chi : A_+ \to \mathbb{C}$ if there is a vector $v \in V$ such that the following properties hold:

- $av = \chi(a)v$ for any $a \in A_+$;
- v generates V as a $\mathbb{C}[G]_q$-module.

Also, in this case we say that v is a *highest weight vector* and χ is the *highest weight* of V.

(2) We say that a $\mathbb{C}[G]_q$-module V is a *Verma module* with *highest weight* $\chi : A_+ \to \mathbb{C}$ if
$$V \simeq \mathbb{C}[G]_q \otimes_{A_+} \mathbb{C}_\chi$$
as a $\mathbb{C}[G]_q$-module. Here we assume that $\mathbb{C}[G]_q$ acts on itself by left multiplication and A_+ acts on $\mathbb{C}[G]_q$ by right multiplication. In that case we write $V = V(\chi)$.

REMARK 5.1.2. It is clear that the Verma module $V(\chi)$ is a highest weight module with a highest weight vector $1 \otimes \mathbf{1}_\chi$ (where 1 is the unit element in $\mathbb{C}[G]_q$) and the highest weight χ. Of course, the highest weight is unique, while a highest weight vector is unique up to multiplication by a constant.

The following exercise utilizes the same machinery as is usually used in the theory of Verma modules over the universal enveloping algebras. This is possible because of the "triangular decomposition" theorem for $\mathbb{C}[G]_q$ (see Theorem 2.2.1).

EXERCISE 5.1.3. Prove the following statements:
(1) $V(\chi)$ is a nontrivial $\mathbb{C}[G]_q$-module if and only if $\chi(A_{++}) \neq 0$.
(2) $V(\chi)$ has a unique simple quotient, denoted by $F(\chi)$.
(3) Every simple finite-dimensional $\mathbb{C}[G]_q$-module is of the form $F(\chi)$. Moreover, it is always one-dimensional.

Let $\Lambda \in P_+$ be a dominant weight, and $L(\Lambda)$ the simple finite-dimensional $U_q\mathfrak{g}$-module with the highest weight Λ. Given an element w of the Weyl group W of $\mathfrak{g}$, we consider a two-sided $*$-invariant ideal $I_w \subset \mathbb{C}[G]_q$ generated by $c_{l,\Lambda}^\Lambda$ such that

(5.1.1) $$l\left(U_q\mathfrak{b}_+ L(\Lambda)_{w\Lambda}\right) = 0.$$

Denote by $c_{w\Lambda}$ the image of $c_{-w\Lambda,\Lambda}^\Lambda$ in $\mathbb{C}[G]_q/I_w$.

For the rest of the section, we assume that $0 < q < 1$ (however, this condition can be weakened in many cases). Also, we will use the notation $\langle \cdot, \cdot \rangle$ for the scalar product in a unitarizable $\mathbb{C}[G]_q$-module and $P(\Lambda)$ for the set of weights of the simple finite-dimensional $U_q\mathfrak{g}$-module $L(\Lambda)$.

Given a simple $\mathbb{C}[G]_q$-module V, the kernel of the representation of $\mathbb{C}[G]_q$ in V is a primitive ideal in $\mathbb{C}[G]_q$. On the other hand, it is easy to see that any primitive ideal $\mathcal{P}$ in $\mathbb{C}[G]_q$ defines a simple $\mathbb{C}[G]_q$-module $V = \mathbb{C}[G]_q/\mathcal{P}$ on which

$\mathbb{C}[G]_q$ acts by left multiplications. Therefore, the simple representations of $\mathbb{C}[G]_q$ are in a one-to-one correspondence with the primitive ideals in $\mathbb{C}[G]_q$.

PROPOSITION 5.1.4. *Let V be a unitarizable simple $\mathbb{C}[G]_q$-module, and $\mathcal{P}$ the corresponding primitive ideal in $\mathbb{C}[G]_q$. Then there exists a unique $w \in W$ such that the following holds:*

(1) $\mathcal{P} \supset I_w$;
(2) $\mathcal{P}/I_w$ *contains neither $c_{w\Lambda}$ nor $c^*_{w\Lambda}$ for any dominant weight Λ.*

In particular, to each unitarizable simple $\mathbb{C}[G]_q$-module there corresponds a unique Schubert cell X_w.

PROOF. Fix a dominant weight Λ and consider a partial order on the set of one-dimensional weight spaces of $L(\Lambda)^*$ induced by the following preordering:

$$l \leq l' \quad \text{if} \quad l' \in U_q \mathfrak{b}_+ l,$$

where l and l' are weight vectors in $L(\Lambda)^*$. Consider the set

$$\mathcal{D}(\Lambda) = \{l \in L(\Lambda)^* \mid l \text{ is a weight vector, and } c^\Lambda_{l,\Lambda} \notin P\}.$$

It is easy to see that $\mathcal{D}(\Lambda)$ has a maximal element l with respect to this partial order. Suppose l has weight λ. Assume that there exists $l' \in L(\Lambda)^*$ which is also maximal in $\mathcal{D}(\Lambda)$. Then we have the following relation in $\mathbb{C}[G]_q/\mathcal{P}$:

(5.1.2)
$$c^\Lambda_{l,\Lambda} c^\Lambda_{l',\Lambda} = q^{2((\Lambda,\Lambda)-(\lambda,\lambda'))} c^\Lambda_{l',\Lambda} c^\Lambda_{l,\Lambda} = q^{4((\Lambda,\Lambda)-(\lambda,\lambda'))} c^\Lambda_{l,\Lambda} c^\Lambda_{l',\Lambda}.$$

Since $\mathcal{P}$ is a primitive ideal and q is not a root of unity, we have that (5.1.2) is possible if and only if $(\Lambda, \Lambda) = (\lambda, \lambda')$.

Recall the following well-known result (cf. [**Kac90**, Proposition 11.4]).

LEMMA 5.1.5. *If $\Lambda \in P_+$ and μ is a weight of $L(\Lambda)$, then $(\Lambda, \Lambda) \geq (\lambda, \mu)$, with equality if and only if $\mu = \lambda \in W(\Lambda)$.*

By Lemma 5.1.5, $\lambda = \lambda' \in W(-\Lambda)$. Therefore, there exists a unique (up to multiplication by a scalar) maximal element $l_0 \in \mathcal{D}(\Lambda)$ with the weight $-w_\Lambda \Lambda$ for some $w_\Lambda \in W$. Consider a pair of matrix elements $c^\Lambda_{l_0,\Lambda}$ and $c^{\Lambda'}_{l'_0,\Lambda'}$ which correspond to different dominant weights Λ and Λ' respectively. Repeating the argument given above, we get that $(w_\Lambda \Lambda, w_{\Lambda'} \Lambda') = (\Lambda, \Lambda')$. The following result is also well known.

LEMMA 5.1.6. *Let $\Lambda \mapsto w_\Lambda$ be a function from P_+ to W. The following two conditions are equivalent:*

(1) $(w_\Lambda \Lambda, w_{\Lambda'} \Lambda') = (\Lambda, \Lambda')$ *for any $\Lambda, \Lambda' \in P_+$.*
(2) *There exists a unique $w \in W$ such that $w\Lambda = w_\Lambda \Lambda$ for any $\Lambda \in P_+$.*

Applying Lemma 5.1.6 to the function $\Lambda \mapsto w_\Lambda$ constructed in the course of the proof, we obtain $w \in W$ such that $w\Lambda = w_\Lambda \Lambda$ for any $\Lambda \in P_+$. Hence, for any $\Lambda \in P_+$, $c^\Lambda_{l,\Lambda} \in P$ is equivalent to $l \not\leq l_{-w\Lambda}$. Therefore, $\mathcal{P} \supset I_w$ and $\mathcal{P}/I_w \not\ni c_{w\Lambda}$.

Since both $\mathcal{P}$ and I_w are invariant with respect to the involution $*$, the same holds for $c^*_{w\Lambda}$. The easy proof of the uniqueness of w is left to the reader. The proposition is proved. □

5.2. Primitive ideals and Schubert cells.

In this subsection we study the relation between the primitive ideals in $\mathbb{C}[G]_q$ which correspond to unitarizable simple $\mathbb{C}[G]_q$-modules and the Schubert cells in the flag manifold G/B_+ (cf. Chapter 1, Section 5.2). Recall that the Schubert cells X_w are parameterized by the elements w of the Weyl group W.

DEFINITION 5.2.1. We say that a primitive ideal $\mathcal{P}$ in $\mathbb{C}[G]_q$ corresponds to a Schubert cell X_w ($w \in W$) if $\mathcal{P}$ satisfies both conditions of Proposition 5.1.4. Accordingly, we say that a unitarizable simple $\mathbb{C}[G]_q$-module V corresponds to a Schubert cell X_w if the corresponding primitive ideal $\mathcal{P}$ satisfies both conditions of Proposition 5.1.4.

PROPOSITION 5.2.2. *For any dominant weight $\Lambda \in P_+$, the following identity holds:*

$$\tag{5.2.1} \sum_{\mu,i} \left(c^{\Lambda}_{-\mu,i,\Lambda}\right)^* c^{\Lambda}_{-\mu,i,\Lambda} = 1,$$

where the sum is taken over all weights μ of $L(\Lambda)$ and $i = 1, 2, \ldots, \dim L(\Lambda)$.

PROOF. Let T_Λ be the matrix consisting of the matrix elements $(T_\Lambda)_{ab}$ of $L(\Lambda)$ in the orthonormal basis $\{v^{(i)}_\lambda\}$ (cf. Section 2.3). Let $T_\Lambda \otimes T_\Lambda$ be the matrix with the matrix elements

$$(T_\Lambda \otimes T_\Lambda)_{ij} = \sum_k (T_\Lambda)_{ik} \otimes (T_\Lambda)_{kj}.$$

If $\Delta : \mathbb{C}[G]_q \to \mathbb{C}[G]_q \otimes \mathbb{C}[G]_q$ is the comultiplication and $S : \mathbb{C}[G]_q \to \mathbb{C}[G]_q$ the antipode in $\mathbb{C}[G]_q$, then

$$\tag{5.2.2} \Delta(T_\Lambda) = T_\Lambda \otimes T_\Lambda, \qquad T_\Lambda S(T_\Lambda) = S(T_\Lambda) T_\Lambda = 1.$$

By the definition of the involution $*$ in $\mathbb{C}[G]_q$, one can easily derive the formula

$$\tag{5.2.3} T_\Lambda \left({}^t T_\Lambda\right)^* = \left({}^t T_\Lambda\right)^* T_\Lambda = 1,$$

where ${}^t T_\Lambda$ is the transposed matrix and $(A^*)_{ij} = a^*_{ji}$ for any matrix $A = ((a_{ij}))$. Now, (5.2.1) immediately follows from (5.2.3). □

COROLLARY 5.2.3. *Let V be a unitarizable $\mathbb{C}[G]_q$-module with a scalar product $\langle \cdot, \cdot \rangle$, Λ a dominant weight of $\mathfrak{h}$, and μ a weight of $L(\Lambda)$. Then for any $v \in V$, we have*

$$\tag{5.2.4} \langle c^{\Lambda}_{-\mu,\Lambda} v, c^{\Lambda}_{-\mu,\Lambda} v \rangle \leq \langle v, v \rangle.$$

PROOF. The assertion immediately follows from (5.2.1). □

PROPOSITION 5.2.4. *Let V be a unitarizable simple $\mathbb{C}[G]_q$-module. Suppose that V corresponds to a Schubert cell X_w for some $w \in W$. Then the action of $c^{\Lambda}_{-w\Lambda,\Lambda}$ in V has trivial kernel for any $\Lambda \in P_+$.*

PROOF. It follows from (2.1.3) that $\operatorname{Ker} c^{\Lambda}_{-w\Lambda,\Lambda}$ is an invariant subspace in V. Since V is simple, it must be trivial. □

Suppose that V is a unitarizable simple $\mathbb{C}[G]_q$-module corresponding to a Schubert cell X_w. Then the $*$-algebra $B_w \subset \operatorname{End}_\mathbb{C}(V)$ generated by the elements of the form $c^{\Lambda}_{-w\Lambda,\Lambda}$ ($\Lambda \in P_+$) is commutative by (2.1.3). Let $\gamma : B_w \to \mathbb{C}$ be a homomorphism of commutative algebras with unity. The set of all such homomorphisms is the maximal spectrum of B_w (denoted by $\operatorname{Spec}_m B_w$).

DEFINITION 5.2.5. We say that a nonzero vector subspace $V_\gamma \subset V$ is a *weight space* with *weight* $\gamma \in \mathrm{Spec}_m B_w$ if the following holds:
$$V_\gamma = \left\{ v \in V \mid c^\Lambda_{-w\Lambda,\Lambda} v = \gamma \left(c^\Lambda_{-w\Lambda,\Lambda} \right) v \text{ for any } \Lambda \in P_+ \right\}.$$

PROPOSITION 5.2.6. *Let $V = V(\chi)$ be a unitarizable simple Verma module over $\mathbb{C}[G]_q$ corresponding to a Schubert cell X_w. Then V has a weight decomposition*
$$V = \bigoplus_\gamma V_\gamma, \ \gamma \in \mathrm{Spec}_m B_w.$$

PROOF. Recall that
$$V = \mathbb{C}[G]_q \otimes_{A_+} \mathbb{C}_\chi.$$
Let $v_\chi \in V$ be the image of $1 \otimes \mathbf{1}_\chi$. Then v_χ generates V as a $\mathbb{C}[G]_q$-module. For any $\Lambda \in P_+$, we have
$$c^\Lambda_{-w\Lambda,\Lambda} v_\chi = \chi \left(c^\Lambda_{-w\Lambda,\Lambda} \right) v_\chi.$$
On the other hand, (2.1.3) implies that
$$(5.2.5) \qquad c^\Lambda_{-w\Lambda,\Lambda} f = q^{d_{f,\Lambda}} f c^\Lambda_{-w\Lambda,\Lambda} \bmod I_w,$$
where f is a matrix element corresponding to the weight vectors, and $d_{f,\Lambda}$ an integer.

The statement of the proposition follows from (5.2.5) and the fact that V is spanned by the monomials of the form $f_1 f_2 \cdots f_n v_\chi$, where f_i are matrix elements of the type $c^{w_0 \Lambda}_{\mu, w_0 \Lambda}$. Indeed, the $\mathbb{C}$-subalgebra of B_w generated by $c^\Lambda_{-w\Lambda,\Lambda}$ ($\Lambda \in P_+$) is diagonalizable in an orthonormal basis of V. Then the eigenvectors are also eigenvectors for the elements of the form $\left(c^\Lambda_{-w\Lambda,\Lambda} \right)^*$ ($\Lambda \in P_+$), hence for B_w. We leave the remaining details to the reader. $\square$

Let $V = V(\chi)$ be as in Proposition 5.2.6. We introduce a partial order on the set of weights of $V = \bigoplus_\gamma V_\gamma$. Namely, $\gamma_1 \leq \gamma_2$ if and only if $\left| \gamma_1 \left(c^\Lambda_{-w\Lambda,\Lambda} \right) \right| \leq \left| \gamma_2 \left(c^\Lambda_{-w\Lambda,\Lambda} \right) \right|$ for any $\Lambda \in P_+$. Then it follows from (2.1.3) and (5.2.5) that $\gamma \leq \chi$, where χ is the weight of $v_\chi \in V$. (We keep the same notation for the obvious reason: the homomorphism χ has a canonical extension to B_w such that $\chi((c^\Lambda_{-w\Lambda,\Lambda})^*) = \overline{\chi(c^\Lambda_{-w\Lambda,\Lambda})}$.)

PROPOSITION 5.2.7. *If $v \in V_\chi$, then $c^\Lambda_{-\mu,\Lambda} v = 0$ for any $\Lambda \in P_+$ and $\mu \in P(\Lambda)$.*

PROOF. Suppose that $v \in V_\chi$ and $c^\Lambda_{-\mu,\Lambda} v \neq 0$ for some Λ and μ. Then $c^\Lambda_{-\mu,\Lambda} v$ is a common eigenvector for B_w with the weight
$$\gamma = \chi q^{-((\Lambda,\Lambda)-(w\Lambda,\mu))} > \chi$$
(if $0 < q < 1$). This contradicts the assumption that χ is the highest weight of V. $\square$

COROLLARY 5.2.8. *Let $V = V(\chi)$ be as in Proposition 5.2.6. Then for any $\Lambda \in P_+$, we have that $\left| \chi \left(c^\Lambda_{-w\Lambda,\Lambda} \right) \right| = 1$.*

PROOF. The assertion immediately follows from Propositions 5.2.2 and 5.2.7. $\square$

EXERCISE 5.2.9. Prove that the statements of Proposition 5.2.7 and Corollary 5.2.8 hold with $V(\chi)$ replaced by $F(\chi)$.

5.3. Representations of $\mathbb{C}[K]_q$ and Schubert cells.

We will now consider irreducible representations $\pi : \mathbb{C}[K]_q \to \operatorname{End} H$ in a separable Hilbert space H.

DEFINITION 5.3.1. Suppose that V is a unitarizable $\mathbb{C}[G]_q$-module with a scalar product $\langle , \rangle$. Consider its completion, a separable Hilbert space H. As follows from Theorem 2.2.1 and Proposition 5.2.2, for any $a \in \mathbb{C}[K]_q$ the action of a in V can be extended to a bounded linear operator in H. The resulting representation $\pi : \mathbb{C}[K]_q \to \operatorname{End} H$ is called a *-representation* of $\mathbb{C}[K]_q$ in the separable Hilbert space H.

PROPOSITION 5.3.2. *For any irreducible representation π of $\mathbb{C}[K]_q$, a dominant weight $\Lambda \in P_+$, and $\mu \in P(\Lambda)$, we have that $\|\pi(c^\Lambda_{-\mu,\Lambda})\| \le 1$, where $\|\cdot\|$ is the operator norm.*

PROOF. The assertion immediately follows from (5.2.1). □

Note that while in Sections 5.1 and 5.2 we deal with unitarizable simple $\mathbb{C}[G]_q$-modules, all of the results of Section 5.1, as well as Proposition 5.2.4, hold for irreducible representations of $\mathbb{C}[K]_q$. Let us summarize the most important of them in the following theorem.

THEOREM 5.3.3. (1) *Given an irreducible representation π of $\mathbb{C}[K]_q$ in a separable Hilbert space H, there exists a unique $w \in W$ such that $\operatorname{Ker} \pi \supset I_w$, where I_w is as defined in Section 5.1. Then we say that π corresponds to the Schubert cell X_w.*

(2) *If an irreducible representation $\pi : \mathbb{C}[K]_q \to \operatorname{End} H$ corresponds to a Schubert cell X_w, then the elements of the form $\pi(c^\Lambda_{-w\Lambda,\Lambda})$ ($\Lambda \in P_+$) generate a commutative $*$-algebra $B_w \subset \operatorname{End} H$. In particular, $\pi(c^\Lambda_{-w\Lambda,\Lambda})$ is a normal bounded operator in H:*

$$(5.3.1) \qquad \pi(c^\Lambda_{-w\Lambda,\Lambda})^* \pi(c^\Lambda_{-w\Lambda,\Lambda}) = \pi(c^\Lambda_{-w\Lambda,\Lambda}) \pi(c^\Lambda_{-w\Lambda,\Lambda})^*.$$

(3) *The kernel of $\pi(c^\Lambda_{-w\Lambda,\Lambda})$ is trivial for any $\Lambda \in P_+$.*

PROPOSITION 5.3.4. *The spectrum of $\pi(c^\Lambda_{-w\Lambda,\Lambda}) \in \operatorname{End} H$ is the union of zero and a set E which is discrete in $\mathbb{C} \setminus \{0\}$.*

PROOF. Let $\omega_1, \omega_2, \ldots, \omega_n$ be the dominant weights of $\mathfrak{g}$. Consider the group Γ of dilations in complex line $\mathbb{C}$ generated by the numbers $q^{(\Lambda,\omega_i)-(w\Lambda,\mu)}$, where $\mu \in P(\omega_i)$, $i = 1, 2, \ldots, n$. We need two auxiliary lemmas.

LEMMA 5.3.5. *Given a nonzero complex number z_0, a dominant weight $\Lambda \in P_+$, and an element w of the Weyl group W, the set $\{q^{(\Lambda,\mu_1)-(w\Lambda,\mu_2)} z_0\}$ is discrete in $\mathbb{C} \setminus \{0\}$, where μ_1, μ_2 run through the integer weight lattice of $\mathfrak{g}$.*

PROOF. Indeed, consider the set $\{a_{\mu_1,\mu_2}\}$, where $a_{\mu_1,\mu_2} = (\Lambda, \mu_1 - w^{-1}\mu_2)$, and μ_1, μ_2 are as above. Then we have that

$$a_{\mu_1,\mu_2} - a_{\mu_3,\mu_4} = \psi\left(\mu_1 - \mu_2 + w^{-1}(\mu_4 - \mu_3)\right),$$

where $\psi : \gamma \mapsto (\Lambda, \gamma)$. Since

$$\psi(\gamma) = \sum_{i=1}^{n} m_i p_i,$$

where $p_i = (\Lambda, \omega_i)$ are nonnegative rational numbers and m_i are integers, the image of the weight lattice with respect to the map ψ is discrete in $\mathbb{C} \setminus \{0\}$. Therefore, the above-mentioned set $\{a_{\mu_1,\mu_2}\}$ is discrete in $\mathbb{C} \setminus \{0\}$. □

LEMMA 5.3.6. *For any nonzero complex number z_0, there exists an open subset $\Omega_{z_0} \subset \mathbb{C}$ such that the following holds:*
(1) $\Omega_{z_0} \ni z_0$, $\Omega_{z_0} \neq \mathbb{C}$,
(2) Ω_{z_0} *is invariant with respect to the group* Γ.

PROOF. By Lemma 5.3.5, there exists a disk $\Omega \ni z_0$ such that $g\Omega \cap \Omega = \emptyset$ for any $g \in \Gamma, g \neq 1$. Therefore, $\Omega_{z_0} = \bigcup_{g \in \Gamma} g\Omega$ satisfies the statement of the lemma. □

Now we can prove Proposition 5.3.4. Let $z_0 \neq 0$ be a point of the spectrum of $\pi\left(c_{-w\Lambda,\Lambda}^{\Lambda}\right)$. Choose an open set $\Omega_{z_0} \ni z_0$ which satisfies Lemma 5.3.6 and denote by $E\left(\Omega_{z_0}\right)$ the spectral projection of the normal operator $\pi\left(c_{-w\Lambda,\Lambda}^{\Lambda}\right)$ corresponding to Ω_{z_0}.

By Theorem 2.2.1, Proposition 2.3.2, and (2.1.3), we get that $E\left(\Omega_{z_0}\right)\pi(f) = \pi(f)E\left(\Omega_{z_0}\right)$ for any $f \in \mathbb{C}[K]_q$. As is well known (cf. [**Lan75**]), the irreducibility of π implies that $E\left(\Omega_{z_0}\right) = c\,\mathrm{id}_H$, where $c \in \mathbb{C}$. By the construction of Ω_{z_0} (see Lemma 5.3.6), the spectrum of $\pi\left(c_{-w\Lambda,\Lambda}^{\Lambda}\right)$ lies in $\Omega_{z_0} \cup \{0\}$. Since the spectrum of $\pi\left(c_{-w\Lambda,\Lambda}^{\Lambda}\right)$ is closed, zero belongs to the spectrum of $\pi\left(c_{-w\Lambda,\Lambda}^{\Lambda}\right)$. Then E is just the intersection of Ω_{z_0} and the spectrum of $\pi\left(c_{-w\Lambda,\Lambda}^{\Lambda}\right)$. The theorem is proved. □

By the spectral theorem and Proposition 5.3.4, there is a decomposition of H into the spectral subspaces

$$H = \bigoplus_{\sigma \in E} H_\sigma(w, \Lambda),$$

where

$$H_\sigma(w, \Lambda) = \left\{v \in H \mid \pi\left(c_{-w\Lambda,\Lambda}^{\Lambda}\right)v = \sigma v\right\}.$$

Introduce a partial order in E by setting $\sigma_1 \geq \sigma_2$ whenever $|\sigma_1| \geq |\sigma_2|$. Let $\sigma_0(w, \Lambda)$ be a maximal element of E with respect to this partial order. Note that if $z_0 \in E$ and $gz = z$ for some $g \in \Gamma$ (see the proof of Proposition 5.3.4), then $g = 1$. Therefore, if $\sigma \in E$ and $\sigma \neq \sigma_0(w, \Lambda)$, then $|\sigma| < |\sigma_0(w, \Lambda)|$.

PROPOSITION 5.3.7. *Every spectral subspace of the form $H_{\sigma_0(w,\Lambda)}(w, \Lambda)$ is one-dimensional.*

PROOF. Consider the action of the commutative $*$-algebra B_w (see Theorem 5.3.3) in $H_{\sigma_0(w,\Lambda)}(w, \Lambda)$. There are two possibilities:

(1) $H_{\sigma_0(w,\Lambda)}(w, \Lambda)$ contains a closed proper B_w-invariant subspace $\hat{H}_{\sigma_0(w,\Lambda)}$.
(2) There is no such subspace.

Consider the case (1). Then $\hat{H}_{\sigma_0(w,\Lambda)}$ generates a Hilbert subspace in H invariant with respect to $\pi(f)$ for any $f \in \mathbb{C}[K]_q$. It is easy to get from (2.1.3) and (5.2.1) that this subspace has a decomposition $\hat{H}_{\sigma_0(w,\Lambda)} \oplus H'$, where

$$H' = \bigoplus_{|\sigma| < |\sigma_0(w,\Lambda)|} H_\sigma(w, \Lambda).$$

Indeed, (2.1.3) and (5.2.1) imply that

- $\pi(c_{-\mu,\lambda}^\lambda)$ acts trivially in $H_{\sigma_0(w,\Lambda)}(w, \Lambda)$ for any $\lambda \in P_+$ and $\mu \in P(\lambda)$ such that $\mu \neq w\lambda$;
- $\pi(c_{-\mu,\lambda}^\lambda)^*(H_{\sigma_0(w,\Lambda)}(w, \Lambda))$ is spanned by eigenvectors of $\pi\left(c_{-w\Lambda,\Lambda}^{\Lambda}\right)$ such that for any eigenvalue λ, we have $|\lambda| < |\sigma_0(w, \Lambda)|$.

Since $\hat{H}_{\sigma_0(w,\Lambda)} \subsetneq H_{\sigma_0(w,\Lambda)}(w,\Lambda)$, we get a contradiction. Hence, (1) is impossible.

Assuming that (2) holds, we see that any operator of the form $\pi(c^\lambda_{-w\lambda,\lambda})$ ($\lambda \in P_+$) acts as $c(\lambda) \cdot \text{id}$ in $H_{\sigma_0(w,\Lambda)}(w,\Lambda)$, where $c(\lambda)$ is a constant. If the dimension of $\mathbb{C}H_{\sigma_0(w,\Lambda)}(w,\Lambda)$ is greater than one, we can find a pair of orthogonal vectors v_1 and v_2 which belong to this space. They generate two $\pi(\mathbb{C}[K]_q)$-invariant orthogonal subspaces in H. Once again, this contradicts the irreducibility of π. Therefore, $\dim_{\mathbb{C}} H_{\sigma_0(w,\Lambda)}(w,\Lambda) = 1$, and the proposition is proved. $\square$

COROLLARY 5.3.8. (1) *The subspace $H_{\sigma_0(w,\Lambda)}(w,\Lambda)$ in H does not depend on $\Lambda \in P_+$.*

(2) *For any $\Lambda \in P_+$, $|\sigma_0(w,\Lambda)| = 1$.*

PROOF. The proof of (2) is the same as that of Corollary 5.2.8. Suppose that for some $\Lambda, \Lambda' \in P_+$, the subspaces $H_{\sigma_0(w,\Lambda)}(w,\Lambda)$ and $H_{\sigma_0(w,\Lambda')}(w,\Lambda')$ in H are distinct. Both of them are B_w-invariant, and, therefore, they generate distinct proper $\pi(\mathbb{C}[K]_q)$-invariant subspaces in H. This contradicts the irreducibility of π. $\square$

Choose a generator $v^+ \neq 0$ of the one-dimensional space $H_0 = H_{\sigma_0(w,\Lambda)}(w,\Lambda)$.

PROPOSITION 5.3.9. (1) *There exists a homomorphism $\chi : A_+ \to \mathbb{C}$ of algebras with unity such that for any $f \in A_+$,*

(5.3.2) $$\pi(f)v^+ = \chi(f)v^+.$$

(2) *v^+ is a highest weight vector of a unitarizable simple $\mathbb{C}[G]_q$-submodule $V \subset H$ with highest weight χ. The closure of V coincides with H.*

PROOF. The first statement follows from Corollary 5.3.8. To prove (2), consider a $\mathbb{C}[G]_q$-submodule V in H generated by v^+. By Theorem 2.2.1 and (2.1.3), V is a unitarizable highest weight module with the highest weight χ. Indeed, any nontrivial unitarizable submodule $V' \subset V$ generates a Hilbert subspace $H' = \bar{V}' \subset H$ invariant with respect to $\pi(\mathbb{C}[K]_q)$. If V' contains H_0, then $V' = V$. If V' does not contain H_0, then $H' \subset H$ is a proper subset, hence π is not irreducible. This contradiction shows that $V' = V$, therefore, V is simple. $\square$

DEFINITION 5.3.10. (1) A homomorphism $\chi : A_+ \to \mathbb{C}$ of algebras with unity is called the *highest weight* of an irreducible representation π of $\mathbb{C}[K]_q$ if for any $f \in A_+$, (5.3.2) holds.

(2) The unitarizable simple $\mathbb{C}[G]_q$-module V from Proposition 5.3.9 is called the *underlying module* for π.

THEOREM 5.3.11. *Two irreducible representations of $\mathbb{C}[K]_q$ are unitary equivalent if and only if their highest weights coincide.*

PROOF. In one direction, the statement is obvious. Indeed, if two irreducible representations are unitary equivalent, their underlying modules are isomorphic, hence the highest weights coincide.

Suppose that we have two irreducible representations π_1 and π_2 of $\mathbb{C}[K]_q$ in separable Hilbert spaces H_1 and H_2 respectively. Suppose that their highest weights are equal (denote them by χ). As we have seen, both H_1 and H_2 are completions of some unitarizable simple highest weight $\mathbb{C}[G]_q$-modules V_1 and V_2 with highest weight vectors v_1 and v_2 respectively.

Consider the $\mathbb{C}[G]_q$-module $V = V_1 \oplus V_2$. Then the line $\mathbb{C}(v_1 + v_2)$ is A_+-invariant and generates a highest weight submodule $N \subset V$. The highest weight of N is obviously equal to χ, and we can choose $v = v_1 + v_2$ as a highest weight vector in N.

Consider the projection $p_1 : N \to V_1$. Then p_1 is a morphism of $\mathbb{C}[G]_q$-modules, and $p_1(v) = v_1$. Since v_1 generates V_1, p_1 is an epimorphism. Consider the kernel of p_1, which is equal to $V_2 \cap N$. This is a $\mathbb{C}[G]_q$-submodule in N. Since v_2 is the only (up to a scalar) vector in V_2 of weight χ, $v_2 \notin V_2 \cap N$. But V_2 is simple, and $V_2 \supsetneq V_2 \cap N$. Therefore, $V_2 \cap N = 0$, and p_1 is an isomorphism.

Similarly, the projection $p_2 : N \to V_2$ is an isomorphism. Therefore, $V_1 \simeq V_2$ and $H_1 \simeq H_2$. □

REMARK 5.3.12. The above proof repeats almost word for word the proof of Theorem 1(c), Part III, Chapter VII from [**Ser87**].

6. Representations of $\mathbb{C}[K]_q$ and symplectic leaves

In the previous section we obtained a description of unitarizable simple $\mathbb{C}[G]_q$-modules and irreducible $*$-representations of $\mathbb{C}[K]_q$ in terms of their highest weights. In the present section we obtain a one-to-one correspondence between the irreducible $*$-representations of $\mathbb{C}[K]_q$ and the symplectic leaves in K. By Theorem 5.2.2 and Lemma 5.2.5 of Chapter 1, any symplectic leaf in K is a product of elementary symplectic leaves corresponding to the vertices of the Dynkin diagram. The construction of irreducible $*$-representations that we describe in this section follows the same pattern.

Throughout this section we continue to assume that $0 < q < 1$.

6.1. Elementary representations of $\mathbb{C}[K]_q$. Let i be a vertex of the Dynkin diagram of $\mathfrak{g}$. Define $\varphi_i : U_{q_i}\mathfrak{sl}(2,\mathbb{C}) \to U_q\mathfrak{g}$ as the canonical embedding of Hopf $*$-algebras given by $X^\pm \mapsto X_i^\pm$, $K \mapsto K_i$. Consider the dual epimorphism $\varphi_i^* : \mathbb{C}[G]_q \to \mathbb{C}[SL_2(\mathbb{C})]_{q_i}$. We see that given an irreducible representation π of $\mathbb{C}[SL_2(\mathbb{C})]_{q_i}$, we get an irreducible representation $\pi \circ \varphi_i^*$ of $\mathbb{C}[G]_q$.

Note that φ_i^* is a Hopf $*$-algebra epimorphism from $\mathbb{C}[K]_q$ onto $\mathbb{C}[SU(2)]_{q_i}$. In particular, any irreducible $*$-representation π of $\mathbb{C}[SU(2)]_{q_i}$ yields an irreducible $*$-representation $\pi \circ \varphi_i^*$ of $\mathbb{C}[K]_q$.

DEFINITION 6.1.1. An irreducible $*$-representation of $\mathbb{C}[K]_q$ is called *elementary* if it is of the form $\pi \circ \varphi_i^*$, where π is an irreducible $*$-representation of $\mathbb{C}[SU(2)]_{q_i}$ and φ_i^* is the canonical epimorphism described above. The corresponding unitarizable simple $\mathbb{C}[G]_q$-module is also called *elementary*.

Recall that we have a one-to-one correspondence between the irreducible $*$-representations of $\mathbb{C}[SU(2)]_{q_i}$ and the symplectic leaves in $SU(2)$ (cf. Remark 4.2.5). This correspondence is a manifestation of a general phenomenon which can be illustrated by the following untuitive observation. In particular, we will see that the setting of the usual orbit method can be incorporated in the same framework.

Let A be an associative algebra over $\mathbb{C}[[h]]$ such that $A_0 = A/hA$ is commutative. We know that in this case A_0 has a canonical Poisson algebra structure with the Poisson bracket given by

$$\{a \bmod h, b \bmod h\} = h^{-1}(ab - ba) \bmod h.$$

Any two-sided ideal $J \subset A$ defines a Poisson ideal $J_0 = J \bmod h$ in A_0. If J is primitive, then spec J_0 is a minimal Poisson subvariety in spec A_0, hence the closure of a symplectic leaf.

EXAMPLE 6.1.2. Consider an arbitrary finite-dimensional complex simple Lie algebra $\mathfrak{g}$. The polynomial algebra $\mathbb{C}[\mathfrak{g}^*]$ carries a Poisson Hopf algebra structure with the Kirillov–Kostant bracket (cf. (3.2.5), Chapter 1). Its quantization is a family of Hopf algebras A_h ($h \in \mathbb{C}$), where A_h is a subalgebra with unity of $U\mathfrak{g}[h]$ generated by the elements of $\mathfrak{g} \subset U\mathfrak{g}[h]$ and the relation $xy - yx = h[x,y]$, where $[\,,\,]$ is the Lie bracket in $\mathfrak{g}$.

If the above-mentioned scheme works (this may not necessarily be the case, in general), we obtain a correspondence between the primitive ideals in $U\mathfrak{g}$ and the coadjoint orbits in $\mathfrak{g}^*$. This is nothing but the correspondence provided by the usual orbit method.

Recall that $\mathfrak{g}^*$ carries a canonical Lie bialgebra structure (cf. Example 2.1.7 in Chapter 1) which arises from the trivial Lie bialgebra structure on $\mathfrak{g}$. Moreover, that Lie bialgebra structure defines a canonical Poisson Lie group structure on $\mathfrak{g}^*$ which is dual to the trivial Poisson Lie group structure on any Lie group G with the Lie algebra $\mathfrak{g}$ (cf. Examples 3.2.7, 3.2.8, and 3.3.6 of Chapter 1). Thus, we see that we can ask a similar question about any Poisson Lie group, namely whether there is a correspondence between irreducible representations of $\mathbb{C}[G]_q$ and the symplectic leaves in G.

As is well known, the usual orbit method provides a one-to-one correspondence between the primitive ideals in $U\mathfrak{g}$ (and thus, irreducible representations of $U\mathfrak{g}$) and the coadjoint orbits in $\mathfrak{g}^*$ if $\mathfrak{g}$ is a nilpotent Lie algebra or a solvable Lie algebra of type I. Recall that given a standard Poisson Lie group structure on a finite-dimensional complex simple Lie group G, the dual Poisson Lie group G^* is a nilpotent Lie group of type I. This can be seen on the level of Lie bialgebras from Example 2.3.7 of Chapter 1. The same holds for the compact real form K of G.

The above observation explains intuitively why there is a one-to-one correspondence between irreducible $*$-representations of $\mathbb{C}[K]_q$ and symplectic leaves in K. Moreover, we see that we can think of it as an example of a general phenomenon often called *quantum orbit method*. On the other hand, we will see in the next section that this correspondence may fail in the case of twisted Lie bialgebra structures on $\mathfrak{g}$, when a nilpotent Poisson Lie group $K^* \subset G^*$ starts to behave as a wild Lie group due to its "wild" Poisson structure.

Let us return to the correspondence between the irreducible $*$-representations of $\mathbb{C}[SU(2)]_{q_i}$ and the symplectic leaves in $SU(2)$ (provided $SU(2)$ is equipped with the standard Poisson Lie group structure). Let

$$D = S_{-1} = \left\{ \begin{pmatrix} a & -\bar{b} \\ b & \bar{a} \end{pmatrix} \mid b \in \mathbb{R}, b < 0 \right\}$$

be a symplectic leaf from Example 5.2.4 of Chapter 1. Denote by V_D the corresponding unitarizable simple $\mathbb{C}[SL_2(\mathbb{C})]_{q_i}$-module and by π_D the corresponding irreducible $\mathbb{C}[SU(2)]_{q_i}$-representation. It corresponds to $t = -1$ in (4.1.3)–(4.1.7).

Then the epimorphism φ_i^* gives rise to an elementary unitarizable simple $\mathbb{C}[G]_q$-module V_{s_i} and an elementary irreducible $*$-representation π_{s_i} of $\mathbb{C}[K]_q$, so that $\pi_{s_i} = \pi_D \circ \varphi_i^* : \mathbb{C}[K]_q \to \operatorname{End} H$. Here s_i denotes the simple reflection in the Weyl group W of $\mathfrak{g}$ which corresponds to the simple root α_i. This notation can be justified by the fact that, as we will see shortly, the representation π_{s_i} in fact corresponds to the Schubert cell X_{s_i} in the sense of Definition 5.2.1.

PROPOSITION 6.1.3. *Suppose that V is a unitarizable simple $\mathbb{C}[G]_q$-module corresponding to the Schubert cell X_w, and i is a vertex of the Dynkin diagram such that $l(s_i w) > l(w)$ (where $l(w)$ stands for the length of $w \in W$). Then the ideal $I_{s_i w} \subset \mathbb{C}[G]_q$ acts trivially in $V_{s_i} \otimes V$.*

PROOF. Fix a weight $\Lambda \in P_+$. For the comultiplication Δ in $\mathbb{C}[G]_q$, we have

(6.1.1) $$\Delta\left(c^\Lambda_{-\mu,i,\Lambda}\right) = \sum_{\lambda,j} c^\Lambda_{-\mu,k,\lambda,j} \otimes c^\Lambda_{\lambda,j,\Lambda},$$

where, as usual, $c^\Lambda_{-\mu,k,\lambda,j}$ is the matrix element of the form $l^{(k)}_{-\mu}(\pi_\Lambda(a) v^{(j)}_\lambda)$ corresponding to the irreducible representation $\pi_\Lambda : U_q\mathfrak{g} \to \text{End}\, L(\Lambda)$.

Suppose that for some matrix element $c^\Lambda_{-\mu,i,\Lambda}$, there exists a term on the right-hand side of (6.1.1) which acts nontrivially in $V_{s_i} \otimes V$. Then $\varphi_i^*(c^\Lambda_{-\mu,k,\lambda,j}) \neq 0$ for some matrix elements on the left-hand side of (6.1.1). Hence, for any $\xi \in U_{q_i}\mathfrak{sl}(2,\mathbb{C})$, we have that

$$l^{(k)}_{-\mu}\left(\pi_\Lambda(\varphi_i(\xi)) v^{(j)}_\lambda\right) \neq 0.$$

Therefore, there exists a nonnegative integer p such that either one or the other of the following two properties holds:

$$\begin{aligned} (X_i^+)^p v^{(j)}_\lambda &= \text{const} \cdot v^{(k)}_\mu \neq 0, \\ (X_i^-)^p v^{(j)}_\lambda &= \text{const} \cdot v^{(k)}_\mu \neq 0. \end{aligned}$$

Let $E_w(\Lambda) = (U_q\mathfrak{b}_+) v_{w\Lambda}$ be the $U_q\mathfrak{b}_+$-module generated by $v_{w\Lambda}$. The following lemma is well known (cf., e.g., [**BGG73**]).

LEMMA 6.1.4. *If $l(s_i w) > l(w)$, then $E_{s_i w}(\Lambda)$ is invariant with respect to the action of $X_i^\pm$.*

Since $v^{(j)}_\lambda \in E_w(\Lambda) \subset E_{s_i w}(\Lambda)$, it follows from the lemma that $v^{(k)}_\mu \in E_{s_i w}(\Lambda)$. But $v^{(k)}_\mu \notin E_{s_i w}(\Lambda)$ by assumption. This contradiction shows that all terms on the right-hand side of (6.1.1) act trivially in $V_{s_i} \otimes V$. $\square$

PROPOSITION 6.1.5. *Under the conditions of Proposition 6.1.3, all matrix elements of the form $c^\Lambda_{-s_i w\Lambda,\Lambda}$ ($\Lambda \in P_+$) act nontrivially in $V_{s_i} \otimes V$.*

PROOF. Similarly to (6.1.1), we can write

(6.1.2) $$\Delta\left(c^\Lambda_{-s_i w\Lambda,\Lambda}\right) = \sum_{\lambda,j} c^\Lambda_{-s_i w\Lambda,\lambda,j} \otimes c^\Lambda_{-\lambda,j,\Lambda}.$$

Since $l(s_i w) > l(w)$, we have $X_i^+ v_{w\Lambda} = 0$. Hence, $v_{w\Lambda}$ is a highest weight vector for the irreducible representation $\pi_\Lambda \circ \varphi_i : U_{q_i}\mathfrak{sl}(2,\mathbb{C}) \to \text{End}\, L(\Lambda)$. Also, $v_{s_i w\Lambda}$ is a lowest weight vector for that representation.

First, we prove that the term $c^\Lambda_{-s_i w\Lambda,w\Lambda} \otimes c^\Lambda_{-w\Lambda,\Lambda}$ on the right-hand side of (6.1.2) acts nontrivially in $V_{s_i} \otimes V$. Note that $\varphi_i^*\left(c^\Lambda_{-s_i w\Lambda,w\Lambda}\right) = a^m_{0,m}$, where $m = (w\Lambda)(H_i)$ and $a^m_{0,m}$ is the matrix element of the standard $m+1$-dimensional representation ρ_m of $U_{q_i}\mathfrak{sl}(2,\mathbb{C})$ ($m = 0,1,2,\ldots$) corresponding to the highest weight vector e_0 (cf. Example 2.3.3). It is easy to see that $a^m_{0,m} = \left(a^1_{0,1}\right)^{2n}$ for any nonnegative integer n. Therefore, $c^\Lambda_{-s_i w\Lambda,w\Lambda} \otimes c^\Lambda_{-w\Lambda,\Lambda}$ acts nontrivially in $V_{s_i} \otimes V$.

Second, we will prove that all the other summands on the right-hand side of (6.1.2) act trivially in $V_{s_i} \otimes V$, which proves the proposition. Indeed, suppose that there is another such term on the right-hand side of (6.1.2) which acts nontrivially in $V_{s_i} \otimes V$. For such a term (and the corresponding values of λ and j), there exists a nonnegative integer p such that

$$v_\lambda^{(j)} = \text{const} \cdot (X_i^+)^p v_{s_i w \Lambda} \neq 0.$$

Therefore, $\lambda = p\alpha_i + w\Lambda - (w\Lambda)(H_i)\alpha_i$. This implies that $p \geq (w\Lambda)(H_i)$. But we have that

$$(X_i^+)^{(w\Lambda)(H_i)} v_{s_i w \Lambda} = \text{const}_1 v_{w\Lambda} \neq 0.$$

If $p > (w\Lambda)(H_i)$, then $(X_i^+)^p v_{s_i w \Lambda} = 0$. Therefore, all terms on the right-hand side of (6.1.2) except $c_{-s_i w\Lambda, w\Lambda}^\Lambda \otimes c_{-w\Lambda, \Lambda}^\Lambda$ disappear. The proposition is proved. □

6.2. Tensor product theorem. Now we construct unitarizable simple modules over $\mathbb{C}[G]_q$ (and irreducible ∗-representations of $\mathbb{C}[K]_q$), using the elementary ones as the blocks of our construction. Then we prove that we get all the representations this way.

THEOREM 6.2.1. *Let $w \in W$ be an element of the Weyl group ($w \neq 1$), and $w = s_{i_1} s_{i_2} \cdots s_{i_k}$ a reduced decomposition of w. Then the $\mathbb{C}[G]_q$-module*

$$V_w = V_{s_{i_1}} \otimes V_{s_{i_2}} \otimes \cdots \otimes V_{i_{s_k}}$$

is unitarizable and simple. It corresponds to the Schubert cell X_w and does not depend (up to isomorphism) on the choice of reduced decomposition for w.

PROOF. It is clear that V_w is unitarizable, simply because any of the elementary modules $V_{s_{i_p}}$ is ($p = 1, 2, \ldots, k$). We prove by induction on $l(w)$ that V_w is simple.

If $l(w) = 1$, then $V_w = V_{s_i}$ for some vertex i, so that the statement of the theorem is true. Suppose that V_w is simple and independent of the choice of a reduced decomposition for any $w \in W$ with $l(w) \leq n$. We will prove that if $l(s_i w) > l(w)$, then $V_{s_i} \otimes V_w$ is a simple module. The proof consists of two steps.

Step 1. Let $\{e_k\}_{k=0}^\infty$ be an orthonormal basis of V_{s_i} described in Proposition 4.1.1. Then there is an orthonormal basis of $V_{s_i} \otimes V_w$ of the form $\{e_k \otimes f_M\}_{k,M}$, where $k \in \mathbb{Z}_+$ and $M \in \mathbb{Z}_+^{l(w)}$. It follows from (6.1.2) and the arguments in the proof of Proposition 6.1.5 that

$$(6.2.1) \qquad c_{-s_i w\Lambda, \Lambda}^\Lambda (e_k \otimes f_M) = \varphi_i^* \left(c_{-s_i w\Lambda, w\Lambda}^\Lambda \right)(e_k) \otimes c_{-w\Lambda, \Lambda}^\Lambda(f_M)$$

for any $\Lambda \in P_+$ and $(k, M) \in \mathbb{Z}_+^{l(w)+1}$. By (6.2.1) and Propositions 6.1.3, 6.1.5, we have that:

- the ∗-algebra $B_{s_i w}$ generated by $c_{-s_i w\Lambda, \Lambda}^\Lambda$ ($\Lambda \in P_+$) (cf. Proposition 5.2.4) is commutative and its action is diagonalizable in the basis $\{e_k \otimes f_M\}$;
- the eigenvalue $\gamma_\Lambda^{(k,M)}$ of $c_{-s_i w\Lambda, \Lambda}^\Lambda$ corresponding to $e_k \otimes f_M$ satisfies $|\gamma_\Lambda^{(k,M)}| < 1$ unless $k = 0$, $M = M_0$, where $M_0 = (0, 0, \ldots, 0)$;
- e_{0, M_0} is the only eigenvector for any $c_{-s_i w\Lambda, \Lambda}^\Lambda$ with the property that $|\gamma_\Lambda^{(0, M_0)}| = 1$. Also, all the eigenvalues $\gamma_\Lambda^{(k, M)}$ are nonzero.

Step 2. Let $V' \subsetneq V_{s_i} \otimes V$ be a $\mathbb{C}[G]_q$-invariant subspace. Then the action of $B_{s_i w}|_{V'}$ is diagonalizable. This can be proved similarly to the proof of Claim 4.1.2. By (2.1.3) and (5.2.4), there exists a vector $v' \in V'$ such that $av' = \gamma(a)v'$ for any

$a \in A_+$. This vector can be chosen among the vectors of the basis $\{e_k \otimes f_M\}$ with $(k, M) \neq (0, M_0)$ (if this is not the case, take the orthogonal complement to V' which is also a $\mathbb{C}[G]_q$-module). Therefore, $\left|\gamma\left(c^\Lambda_{-s_i w\Lambda, \Lambda}\right)\right| < 1$ for some $\Lambda \in P_+$.

Let us call a vector $v' \in V'$ with the above property *primitive* and the corresponding homomorphism $\Gamma : A_+ \to \mathbb{C}$ the *weight* of v'. There is a natural partial order on the set of weights of primitive vectors: $\gamma_1 \prec \gamma_2$ if and only if $|\gamma_1(a)| \geq |\gamma_2(a)|$ for any $a \in \left\{c^\Lambda_{-s_i w\Lambda, \Lambda}\right\}_{\Lambda \in P_+}$.

Let $\gamma_{\max}$ be a maximal element with respect to this order, and $v_{\max}$ the corresponding eigenvector. Then we can prove, using (2.1.3), that $c^\Lambda_{-\mu,\Lambda} v_{\max} = 0$ for any $\mu \neq s_i w\Lambda$. Indeed, any vector of the form $c^\Lambda_{-\mu,\Lambda} v_{\max}$ would be an eigenvector for $B_{s_i w}$ with the weight larger than $\gamma_{\max}$, which is impossible.

It follows from (5.2.1) that

$$(6.2.2) \qquad \sum_{\mu,i} \left(c^\Lambda_{-\mu,i,\Lambda}\right)^* c^\Lambda_{\mu,i,\Lambda} v_{\max} = v_{\max}$$

for a given $\Lambda \in P_+$. Since all except one summands in (6.2.2) disappear, we get

$$(6.2.3) \qquad \left(c^\Lambda_{-s_i w\Lambda, \Lambda}\right)^* c^\Lambda_{-s_i w\Lambda, \Lambda} v_{\max} = v_{\max}.$$

We can assume that $\langle v_{\max}, v_{\max} \rangle = 1$, where $\langle\,,\,\rangle$ is the scalar product in $V_{s_i} \otimes V$. Thus, (6.2.3) shows that $\left|\gamma\left(c^\Lambda_{-s_i w\Lambda, \Lambda}\right)\right| = 1$ for any $\Lambda \in P_+$. But, as we have seen, this is impossible unless $v_{\max} = e_0 \otimes f_{M_0}$. Since we assumed that $e_0 \otimes f_{M_0} \notin V'$, we come to a contradiction. Therefore, $V_{s_i} \otimes V$ is simple.

Propositions 6.1.3 and 6.1.5 imply that $V_{s_i} \otimes V$ corresponds to the Schubert cell $X_{s_i w}$. The fact that it does not depend on the choice of the reduced decomposition of $s_i w$ can be deduced from the induction assumption, and (6.2.1) by comparing the highest weights. We give a direct proof in the next Proposition. $\square$

PROPOSITION 6.2.2. *Suppose that i and j are two vertices of the Dynkin diagram of $\mathfrak{g}$, and m_{ij} is the order of the element $s_i s_j$ in the Weyl group. Then the $\mathbb{C}[G]_q$-module $V_{s_i} \otimes V_{s_j} \otimes V_{s_i} \otimes \cdots$ (m_{ij} times) is isomorphic to $V_{s_j} \otimes V_{s_i} \otimes V_{s_i} \otimes \cdots$ (m_{ij} times).*

PROOF. We already know that those modules are simple and correspond to the same Scubert cell $X_{s_i s_j s_i \cdots} = X_{s_j s_i s_j \cdots}$ (indeed, $s_i s_j s_i \cdots = s_j s_i s_j \cdots$ in W). Hence, it suffices to compare their highest weights.

For an arbitrary $\Lambda \in P_+$ and a reduced decomposition $w = s_{i_1} s_{i_2} \ldots s_{i_k}$, we have

$$(6.2.4) \qquad c^\Lambda_{-w\Lambda, \Lambda}|_{V_{s_{i_1}} \otimes V_{s_{i_2}} \otimes \cdots \otimes V_{s_{i_k}}}$$
$$= \left(a^1_{0,1}\right)^{(s_{i_2} \cdots s_{i_k} \Lambda)(H_{i_1})}$$
$$\otimes \left(a^1_{0,1}\right)^{(s_{i_3} \cdots s_{i_k} \Lambda)(H_{i_2})} \otimes \cdots \otimes \left(a^1_{0,1}\right)^{\Lambda(H_{i_k})},$$

where $a^m_{0,m}$ is the same matrix element as in the proof of Proposition 6.1.5. Therefore,

$$(6.2.5) \qquad \langle c^\Lambda_{-w\Lambda, \Lambda} e_0^{\otimes k}, e_0^{\otimes k} \rangle = (-1)^{p_\Lambda(s_{i_1}, s_{i_2}, \ldots, s_{i_k})},$$

where

$$p_\Lambda(s_{i_1}, \ldots, s_{i_k}) = \Lambda(H_{i_k}) + (s_{i_k} \Lambda)(H_{i_{k-1}}) + \cdots + (s_{i_2} \cdots s_{i_k} \Lambda)(H_{i_1}).$$

It suffices to verify that

(6.2.6) $$p_\Lambda(s_i, s_j, s_i, \ldots) = p_\Lambda(s_j, s_i, s_j, \ldots) \quad (m_{ij} \text{ times}).$$

Let $A = ((a_{ij}))$ be the Cartan matrix of $\mathfrak{g}$. It is well known that m_{ij} is equal to $2, 3, 4, 6$ if $a_{ij}a_{ji}$ is equal to $0, 1, 2, 3$ respectively. Then (6.2.6) can be verified directly for all the four cases. This proves the proposition and completes the proof of Theorem 6.2.1. □

EXERCISE 6.2.3. Prove that any finite-dimensional unitarizable simple $\mathbb{C}[G]_q$-module is one-dimensional. Any such module corresponds to the Schubert cell X_e, where e is the unity in W. They are in one-to-one correspondence with the points of the distinguished maximal torus $T \subset K$ (which consists of zero-dimensional symplectic leaves).

EXERCISE 6.2.4. Let V_τ, $\tau \in T$, be a one-dimensional module from the previous exercise, and V_w ($w \neq e$) a module from Theorem 6.2.1. Show that both $V_\tau \otimes V_w$ and $V_w \otimes V_\tau$ are simple and correspond to the Schubert cell X_w.

EXERCISE 6.2.5. Prove that $V_\tau \otimes V_{\tau'} \simeq V_{\tau\tau'}$ for any $\tau, \tau' \in T$.

PROPOSITION 6.2.6. *Let M_1 and M_2 be two unitarizable simple $\mathbb{C}[G]_q$-modules corresponding to the same Schubert cell X_w. Then there exist $\tau, \tau' \in T$ such that $M_1 \otimes V_\tau \simeq M_2$ and $V_{\tau'} \otimes M_1 \simeq M_2$, where V_τ is a one-dimensional module from Exercise 6.2.4.*

PROOF. Let χ_1 and χ_2 be the highest weights of M_1 and M_2 respectively. By Exercise 6.2.4, $M_1 \otimes V_\tau$ is simple for any $\tau \in T$ and corresponds to the same Schubert cell X_w. Note that both χ_1 and χ_2 are uniquely defined by their values on the matrix elements of the form $c^{\omega_i}_{-w\omega_i, \omega_i}$ ($i = 1, 2, \ldots, n$), where $\{\omega_i\}_{i=1}^n$ is the set of fundamental weights of $\mathfrak{g}$.

It is easy to see from (6.1.1) that

(6.2.7) $$\chi^\tau \left(c^{\omega_i}_{-w\omega_i, \omega_i} \right) = \chi_1 \left(c^{\omega_i}_{-w\omega_i, \omega_i} \right) \tau_i,$$

where $\tau = (\tau_1, \tau_2, \ldots, \tau_n) \in T$ and χ^τ is the highest weight of $M_1 \otimes V_\tau$. Thus, we see that one can find $\tau \in T$ such that $\chi^\tau = \chi_2$. The second statement of the proposition can be proved similarly. □

Now we can summarize the results of this section in one theorem.

THEOREM 6.2.7. (1) *For any unitarizable simple $\mathbb{C}[G]_q$-module V, there exists a unique element $w \in W$ of the Weyl group and a unique element $\tau \in T$ of the distinguished maximal torus such that*

(6.2.8) $$V \simeq V_{s_{i_1}} \otimes V_{s_{i_2}} \otimes \cdots \otimes V_{s_{i_k}} \otimes V_\tau,$$

where $w = s_{i_1} s_{i_2} \cdots s_{i_k}$ is a reduced decomposition of w. The tensor product in (6.2.8) does not depend (up to a unitary equivalence) on the choice of reduced decomposition of w.

(2) Any unitarizable simple $\mathbb{C}[G]_q$-module corresponding to the Schubert cell X_w is of the form (6.2.8), where $w = s_{i_1} s_{i_2} \cdots s_{i_k}$ is a reduced decomposition of w and $\tau \in T$.

(3) Both (1) and (2) hold for irreducible $*$-representations of $\mathbb{C}[K]_q$ in a separable Hilbert space.

COROLLARY 6.2.8. *The irreducible $*$-representations of $\mathbb{C}[K]_q$ are in one-to-one correspondence with the symplectic leaves in K.*

PROOF. By Theorem 5.2.2 of Chapter 1, the symplectic leaves in K are parameterized by the elements of the Weyl group W and by the points of the distinguished maximal torus $\tau \in T$. □

7. Representation theory of the twisted algebras of functions

In the previous section we obtained a quantum analog of Theorem 5.2.2 of Chapter 1 in the form of a parameterization of the irreducible $*$-representations of $\mathbb{C}[K]_q$ by the Schubert cells X_w ($w \in W$) of the flag manifold $G/B_+ \simeq K/T$ and the points $\tau \in T$ of the distinguished maximal torus.

In the present section we consider, instead, the so-called "twisted" quantized algebras of functions, arising from the quantizations of nonstandard Poisson Lie group structures on K. Recall that Theorem 5.1.1 of Chapter 1 describes all quasitriangular Poisson Lie group structures on K. They are parameterized by $a \in \mathbb{R}$ and $u \in \mathfrak{t} \wedge \mathfrak{t}$, where $\mathfrak{t}$ is the Lie algebra of the maximal torus T. We denote them by $K(a, u)$. Note that $K(ta, tu) \simeq K(a, u)$, where $t \in \mathbb{R} \setminus \{0\}$. In particular, the standard Poisson Lie group structure on K is represented by $K(a, 0) \simeq K(1, 0)$.

Most of the proofs are just outlined, as they largely repeat the ideas of the proofs in the nontwisted situation. The careful reader should consider them as exercises. Recall that we assume that $0 < q < 1$.

7.1. Twisted quantized algebras of functions.
Now we will consider the Poisson Lie groups $K^u \simeq K(1, u)$ and $K(0, u)$ and their quantizations. Recall that the quantizations of the co-Poisson Hopf algebras $U\mathfrak{g}^u$ and $U\mathfrak{g}(0, u)$ are the Hopf algebras $U_h\mathfrak{g}^u$ and $U_h\mathfrak{g}(0, u)$, respectively, introduced in Chapter 2.

As was noted in Section 7.4 of Chapter 2, the category of finite-dimensional $U_q\mathfrak{g}^u$-modules is a ribbon category, so that the dual Hopf algebra $\mathbb{C}[G^u]_q$ of the matrix elements of the finite-dimensional representations of $U_q\mathfrak{g}^u$ is well defined. The same holds for the Hopf algebra $\mathbb{C}[G(0, u)]_q$ of the matrix elements of finite-dimensional representations of $U_q\mathfrak{g}(0, u)$.

REMARK 7.1.1. Since $U_q\mathfrak{g}$, $U_q\mathfrak{g}^u$, and $U_q\mathfrak{g}(0, u)$ are all isomorphic as algebras, we can identify $\mathbb{C}[G]_q$, $\mathbb{C}[G^u]_q$, and $\mathbb{C}[G(0, u)]_q$ as vector spaces so that the notation $c_{l,v}^\pi$ stands for the same matrix element of the same representation in any of the above three quantized algebras of functions.

It is easy to see that the same involutive algebra antiautomorphism θ that defines the Hopf $*$-algebra structure on $U_q\mathfrak{g}$ in Proposition 1.2.3 defines also the Hopf $*$-algebra structures on $U_q\mathfrak{g}^u$ and $U_q\mathfrak{g}(0, u)$ respectively. Indeed, the twisting does not change the algebra structure, while the comultiplication is twisted by the conjugation.

DEFINITION 7.1.2. The Hopf $*$-algebra $\mathbb{C}[K^u]_q$ (resp. $\mathbb{C}[K(0, u)]_q$) which is dual to the Hopf $*$-algebra $(U_q\mathfrak{g}^u, \theta)$ (resp. $(U_q\mathfrak{g}(0, u), \theta)$) is called the *quantized algebra of regular functions* on K^u (resp. $K(0, u)$).

Sometimes we call both $\mathbb{C}[K^u]_q$ and $\mathbb{C}[K(0, u)]_q$ *twisted quantized algebras of functions* to distinguish them from the "*standard*" quantization $\mathbb{C}[K]_q$.

The notions of unitarizable $\mathbb{C}[G^u]_q$-module (resp. $\mathbb{C}[G(0,u)]_q$-module) and $*$-representation of $\mathbb{C}[K^u]_q$ (resp. $\mathbb{C}[K(0,u)]_q$) in a separable Hilbert space can be defined similarly to the notions of unitarizable $\mathbb{C}[G]_q$-module and $*$-representation of $\mathbb{C}[K]_q$; see Definitions 1.1.6 and 5.3.1. Contrary to the way we developed the representation theory of $\mathbb{C}[K]_q$, we will present the theory of irreducible $*$-representations of $\mathbb{C}[K^u]_q$ and $\mathbb{C}[K(0,u)]_q$ to emphasize the role of the so-called "quantum tori." We leave it as a simple exercise to reformulate the theory for the corresponding unitarizable modules.

7.2. Representations of $\mathbb{C}[K^u]_q$ and quantum tori. We consider first the case of $\mathbb{C}[K^u]_q$. Recall that $u \in \mathfrak{t} \wedge \mathfrak{t} \subset \mathfrak{h} \wedge \mathfrak{h}$ defines a linear operator $\check{u}: \mathfrak{h} \to \mathfrak{h}^*$. Consider the bilinear form $(\,,\,)_u$ on $\mathfrak{h}$ given by

$$(x,y)_u = (\check{u}x, y).$$

Then the multiplication in $\mathbb{C}[K^u]_q$ gets twisted so that the following version of (2.1.3) holds.

PROPOSITION 7.2.1. *Let $\Lambda, \Lambda' \in P_+$ and $l \in L(\Lambda)^*_{-\lambda}$, $l' \in L(\Lambda')_{-\lambda'}$. Suppose that $v' \in L(\Lambda')_\gamma$ and that $v_\Lambda \in L(\Lambda)_\Lambda$ is a highest vector. Then there exist $a_\nu \in \mathbb{C}$, $l_\nu \in ((U_q\mathfrak{b}_+)l)_{-\lambda+\nu}$, and $l'_\nu \in ((U_q\mathfrak{b}_-)l')_{-\lambda'-\nu}$ such that*

(7.2.1)
$$c^{\Lambda'}_{l',v'} c^{\lambda}_{l,v_\Lambda} = q^{2((\Lambda,\gamma)-(\lambda,\lambda')-i(\Lambda,\gamma)_u+i(\lambda,\lambda')_u)} c^{\Lambda}_{l,v_\Lambda} c^{\Lambda'}_{l',v'} + q^{2(\Lambda,\gamma)} \sum_{\nu \in Q_+} a_\nu c^{\Lambda}_{l_\nu,v_\Lambda} c^{\Lambda'}_{l'_\nu,v'},$$

where Q_+ is the lattice of positive roots of $\mathfrak{g}$.

PROOF. The proof repeats the proof of Proposition 2.1.4, with the only modification that one should use the twisted version of the quasi-R-matrix. It can be easily derived from Section 2.4 of Chapter 2. □

DEFINITION 7.2.2. We say that an irreducible $*$-representation $\pi: \mathbb{C}[K^u]_q \to \operatorname{End} H$ corresponds to a Schubert cell X_w ($w \in W$) if $\operatorname{Ker} \pi \supset I_w$ and $\pi(c^\Lambda_{-w\Lambda,\Lambda}): H \to H$ is a nonzero operator for any $\Lambda \in P_+$.

PROPOSITION 7.2.3. *Every irreducible $*$-representation π of $\mathbb{C}[K^u]_q$ in a separable Hilbert space corresponds to a unique Schubert cell X_w.*

PROOF. The proof repeats the proof of Proposition 5.1.4. □

COROLLARY 7.2.4. *Let $\pi: \mathbb{C}[K^u]_q \to \operatorname{End} H$ be an irreducible $*$-representation of $\mathbb{C}[K^u]_q$ corresponding to a Schubert cell X_w. Then for any $\lambda, \mu \in P_+$, the following relation holds:*

(7.2.2)
$$\pi(c^\lambda_{-w\lambda,\lambda}) \pi(c^\mu_{-w\mu,\mu}) = q^{2i((w\lambda,w\mu)_u - (\lambda,\mu)_u)} \pi(c^{\lambda+\mu}_{-w(\lambda+\mu),\lambda+\mu}).$$

PROOF. The assertion immediately follows from (7.2.1) and can be checked by a straightforward computation. □

For any $w \in W$, introduce the following two functions on $P_+ \times P_+$:

(7.2.3) $\quad\quad\quad\quad \theta_{w,u}(\lambda,\mu) = (w\lambda, w\mu)_u - (\lambda,\mu)_u,$

(7.2.4) $\quad\quad\quad\quad \sigma_{w,u}(\lambda,\mu) = q^{i\theta_{w,u}(\lambda,\mu)}.$

For any irreducible $*$-representation π of $\mathbb{C}[K^u]_q$ in a separable Hilbert space H, define $B_{w,u}$ as a $*$-subalgebra in $\operatorname{End} H$ generated by the linear operators of the form $\pi\left(c^\Lambda_{-w\Lambda,\Lambda}\right)$ ($\Lambda \in P_+$). The following proposition shows that instead of being commutative (as B_w), $B_{w,u}$ satisfies more complicated relations.

PROPOSITION 7.2.5. (1) *The following relations hold in* $\mathbb{C}[G^u]_q$ *for any* $\lambda, \mu \in P_+$:

$$\begin{aligned}
C_\lambda C_\mu &= \sigma_{w,u}(\lambda,\mu) C_{\lambda+\mu}, \\
C_\lambda C_\mu &= (\sigma_{w,u}(\lambda,\mu))^2 C_\mu C_\lambda, \\
C^\lambda_\lambda (C_\mu)^* &= (\sigma_{w,u}(\mu,\lambda))^2 (C_\mu)^* C_\lambda,
\end{aligned}$$

where $C_\lambda = c^\lambda_{-w\lambda,\lambda}$.
(2) *The element*

$$D_\lambda = (C_\lambda)^* C_\lambda = C_\lambda (C_\lambda)^*$$

is central in $B_{w,u}$ *for any* $\lambda \in P_+$.

PROOF. The assertion immediately follows from (7.2.1) and can be checked by a straightforward computation. $\square$

PROPOSITION 7.2.6. *Let* $D_j = D_{\omega_j}$, *where* $\{\omega_j\}_{j=1}^n$ *is the set of fundamental weights. The following relations hold in* $\mathbb{C}[G^u]_q / I_w$ *for any* $\lambda \in P_+$ *and* $\mu \in P(\lambda)$:

$$\begin{aligned}
(7.2.5) \qquad D_j c^\lambda_{-\mu,\lambda} &= q^{-2k_j(\lambda,\mu)} c^\lambda_{-\mu,\lambda}, \\
(7.2.6) \qquad D_j \left(c^\lambda_{-\mu,\lambda}\right)^* &= q^{2k_j(\lambda,\mu)} \left(c^\lambda_{-\mu,\lambda}\right)^*,
\end{aligned}$$

where $k_j(\lambda,\mu) = (\omega_j, \lambda) - (w\omega_j, \mu) \geq 0$.

PROOF. Straightforward computation. $\square$

PROPOSITION 7.2.7. *Let* π *be an irreducible $*$-representation of* $\mathbb{C}[K^u]_q$ *corresponding to a Schubert cell* X_w. *Then* $\pi(D_j)$ *is a selfadjoint bounded operator. Its spectrum is equal to* $\{0\} \cup E_j$, *where* E_j *is a discrete set in* $\mathbb{R} \setminus \{0\}$ *with a unique limit point* 0.

PROOF. The proof repeats the proof of Proposition 5.3.4. $\square$

For any irreducible $*$-representation π of $\mathbb{C}[K^u]_q$ in a separable Hilbert space H, consider the set $E_\pi = E_1 \times E_2 \times \cdots \times E_n$. It is partially ordered so that $\gamma \prec \gamma'$ if and only if $\gamma_j \geq \gamma'_j$ for any $j = 1, 2, \ldots, n$.

DEFINITION 7.2.8. A tuple $\gamma = (\gamma_1, \gamma_2, \ldots, \gamma_n) \in E_\pi$ is called a *weight* of π. A maximal element $\gamma \in E_\pi$ is called a *maximal weight* of π. For any weight $\gamma \in E_\pi$, we define the corresponding *weight space* by

$$H_\gamma = \{v \in H \mid \pi(D_j) v = \gamma_j v \text{ for any } j = 1, 2, \ldots, n\}.$$

Let $\gamma^0 \in E_\pi$ be a maximal weight of π. Consider the corresponding weight space H_{γ^0}. It is obviously contained in any of the eigenspaces $H\left(\gamma^0_j, D_j\right)$ in H which correspond to the maximal eigenvalues γ^0_j of the operators $\pi(D_j)$. The following proposition shows that they coincide.

PROPOSITION 7.2.9. (1) *For any irreducible $*$-representation π of $\mathbb{C}[K^u]_q$ in a separable Hilbert space H and $w \in W$, there is no proper $B_{w,u}$-invariant subspace in $H\left(\gamma_j^0, D_j\right)$ for any $j = 1, 2, \ldots, n$.*

(2) *For any irreducible $*$-representation π of $\mathbb{C}[K^u]_q$ in a separable Hilbert space H, $H_{\gamma^0} = H\left(\gamma_j^0, D_j\right)$ for any $j = 1, 2, \ldots, n$.*

PROOF. Suppose that $\hat{H}$ is a nontrivial subspace in $H\left(\gamma_j^0, D_j\right)$ invariant with respect to $B_{w,u}$. By Proposition 7.2.6,

$$\pi(\mathbb{C}[K^u]_q)\hat{H} \subset \hat{H} \oplus \sum_{\gamma_s \leq \gamma_j^0} H(\gamma_s, D_j),$$

where $H(\gamma_s, D_j)$ is the eigenspace of $\pi(D_j)$ corresponding to the eigenvalue γ_s. Since π is irreducible, $\hat{H} = H\left(\gamma_j^0, D_j\right)$. This proves (1).

In particular, $H\left(\gamma_j^0, D_j\right) = B_{w,u} v$ for any $v \in H\left(\gamma_j^0, D_j\right)$. By Proposition 7.2.5, we have that $H = B_{w,u} v \oplus H_1$, where H_1 denotes the subspace in H spanned by the vectors with weights $\gamma < \gamma_j^0$. Now, let $j_1 \neq j$. Since $H\left(\gamma_j^0, D_j\right)$ is $\pi(D_j)$-invariant, we can assume that $v \in H\left(\gamma_{j_1}^0, D_{j_1}\right)$. Indeed, Proposition 7.2.6 implies that if $v \in H\left(\gamma'_{j_1}, D_{j_1}\right)$, then $H \subset \bigoplus_{\gamma' \leq \gamma_{j_1}} H\left(\gamma', D_{j_1}\right)$. Hence,

$$H\left(\gamma_{j_1}^0, D_{j_1}\right) = B_{w,u} v \subset H\left(\gamma_j^0, D_j\right).$$

Substituting j_1 for j and vice versa, we get

$$H\left(\gamma_j^0, D_j\right) = H\left(\gamma_{j_1}^0, D_{j_1}\right) = H_{\gamma^0}.$$

Proposition 7.2.9 is proved. □

PROPOSITION 7.2.10. *The maximal weight of any irreducible $*$-representation π of $\mathbb{C}[K^u]_q$ is unique and equal to $\gamma^0 = (1, 1, \ldots, 1)$.*

PROOF. The proof is similar to the proof of Proposition 5.3.8 and is based on the use of Propositions 7.2.6, 7.2.9, and the following formula:

(7.2.7) $$\pi(D_\lambda) = \pi(C_\lambda^*) \pi(C_\lambda) = 1 \text{ in } H_{\gamma^0}.$$

Proposition 7.2.10 is proved. □

Recall the definition of quantum tori (cf., for example, [**Rie88**]).

DEFINITION 7.2.11. Let P be an abelian group, and $\kappa : P \times P \to \mathbb{R}$ an antisymmetric bilinear form. Let

$$\sigma(\lambda, \mu) = e^{i\kappa(\lambda, \mu)}.$$

The unital $*$-algebra $A(P, \sigma)$ with the generators U_λ $(\lambda \in P)$ and the relations

(7.2.8) $$U_\lambda^* U_\lambda = U_\lambda U_\lambda^* = 1,$$
(7.2.9) $$U_\lambda U_\mu = \sigma(\lambda, \mu) U_{\lambda + \mu}$$

is called the *quantum torus* corresponding to P and κ.

Consider the subalgebra $P_{w,u}$ in $\mathbb{C}[K^u]_q$ generated by the matrix elements of the form $c^\Lambda_{-w\Lambda, \Lambda}$ ($\Lambda \in P_+$). Let $\mathcal{D}$ be the two-sided $*$-ideal in $P_{w,u}$ generated by $D_j - 1$ ($j = 1, 2, \ldots, n$). It follows from Proposition 7.2.5 and (7.2.7) that an irreducible $*$-representation $\pi : \mathbb{C}[K^u]_q \to \operatorname{End} H$ corresponding to a Schubert cell X_w determines a unique irreducible $*$-representation of the quotient algebra

$P_{w,u}/\mathcal{D}$ in H_{γ^0}. It is easy to see that $P_{w,u}/\mathcal{D}$ is isomorphic to $A(P,\sigma)$, where P is the integer weight lattice and $\sigma = \sigma_{w,u}$ is given by (7.2.3)–(7.2.4).

DEFINITION 7.2.12. The $*$-representation $\chi : A(P,\sigma) \to \operatorname{End} H_{\gamma^0}$ is called the *highest weight* of the irreducible $*$-representation π of $\mathbb{C}[K^u]_q$ in a separable Hilbert space H.

PROPOSITION 7.2.13. *Irreducible $*$-representations of $\mathbb{C}[K^u]_q$ corresponding to the same Schubert cell X_w are unitary equivalent if and only if their highest weights are unitary equivalent.*

PROOF. Similar to the proof of Theorem 5.3.11. □

THEOREM 7.2.14. *For any $w \in W$ and an irreducible $*$-representation χ of the quantum torus $A(P, \sigma_{w,u})$ in a separable Hilbert space H_χ, there exists an irreducible $*$-representation π of $\mathbb{C}[K^u]_q$ which corresponds to the Schubert cell X_w and has the highest weight χ.*

PROOF. Let $\tilde{P} = P \oplus P$ and
$$\tilde{\sigma}((\lambda,\mu),(\lambda',\mu')) = q^{i((\mu,\mu')_u - (\lambda,\lambda')_u)}.$$

They define the quantum torus $A(\tilde{P}, \tilde{\sigma})$. Consider the involution in
$$\mathbb{C}[K^u]_q = \left(\mathbb{C}[K^u]_q\right) \otimes A(\tilde{P}, \tilde{\sigma})$$
given by $(a \otimes b)^* = a^* \otimes b^*$. It is easy to see that it makes $\mathbb{C}[K^u]_q$ a $*$-algebra. Furthermore, consider the $*$-algebra embedding $\iota : \mathbb{C}[K^u]_q \hookrightarrow \mathbb{C}[K^u]_q$ given by
$$\iota : c^\Lambda_{-\mu,i,\lambda,j} \mapsto c^\Lambda_{-\mu,i,\lambda,j} \otimes U_{\lambda,\mu}.$$

We observe that $A(P,\sigma)$ can be embedded into $A(\tilde{P}, \tilde{\sigma})$ as the isomorphic subalgebra $A(P',\sigma')$, where

(7.2.10) $\qquad P' = \{(-w,\lambda,\lambda) \mid \lambda \in P\}, \quad \sigma' = \tilde{\sigma}|_{P' \times P'}.$

LEMMA 7.2.15. *Let $P' \subset \tilde{P}$ be a subgroup of an abelian group $\tilde{P}$. Let $A(\tilde{P}, \tilde{\sigma})$ and $A(P',\sigma')$ be quantum tori such that $\sigma' = \tilde{\sigma}|_{P' \times P'}$. Then for any $*$-representation χ of $A(P',\sigma')$ in a separable Hilbert space H_χ, there exists a separable Hilbert space $H_{\tilde{\chi}} \subset H_\chi$ and a $*$-representation $\tilde{\chi}$ of $A(\tilde{P}, \tilde{\sigma})$ in $H_{\tilde{\chi}}$ such that $\tilde{\chi}(f) = \chi(f)$ for any $f \in A(P',\sigma')$.*

PROOF. Let P'' be a subgroup in $\tilde{P}$ such that $\tilde{P} = P' \oplus P''$. Note that
$$H = A(\tilde{P}, \tilde{\sigma}) \otimes_{A(P',\sigma')} H_\chi$$
has a decomposition $H = \bigoplus_{\mu \in P''} H_\mu$, where $H_\mu = \{U_\mu \otimes x \mid x \in H_\chi\}$. Introduce the scalar product in H as follows:

(7.2.11) $\qquad \langle U_\lambda \otimes x, U_\mu \otimes y \rangle = \delta_{\lambda\mu} \langle x, y \rangle$

for any $\lambda, \mu \in P''$. Consider the completion $\bar{H}$ of H with respect to the scalar product (7.2.11). Then the natural action of $A(\tilde{P}, \tilde{\sigma})$ in $\bar{H}$ yields the $*$-representation $\tilde{\chi}$ we are looking for. □

Now we apply the lemma to the subgroup P' in P described by (7.2.10). Thus, we get a $*$-representation $\tilde{\chi} : A(\tilde{P}, \tilde{\sigma}) \to \operatorname{End} H_{\tilde{\chi}}$ such that $H_\chi \subset H_{\tilde{\chi}}$ and $\tilde{\chi}(f) = \chi(f)$ for any $f \in A(P, \sigma) \simeq A(P', \sigma')$.

Let π_w be an irreducible $*$-representation of $\mathbb{C}[K]_q$ in a separable Hilbert space H corresponding to the Schubert cell X_w. Then $\tilde{\pi} = \pi \otimes \tilde{\chi}$ is a $*$-representation of $\mathbb{C}[K^u]_q$ in $H \otimes H_{\tilde{\chi}}$. Let $l \subset H$ be the unique $\pi(A_+)$-invariant straight line (see Theorem 4.2.1). Let

$$H' = \pi\left(\mathbb{C}[K^u]_q\right)(l \otimes H_{\tilde{\chi}}) \subset H \otimes H_{\tilde{\chi}}.$$

Looking at the weights of vectors in H', it is easy to show that there exists a unique proper $\pi(\mathbb{C}[K^u]_q)$-invariant subspace H'' in H'. The complement is the space of the irreducible $*$-representation of $\mathbb{C}[K^u]_q$ we are looking for. $\square$

7.3. Representations of $\mathbb{C}[K^u]_q$ and symplectic leaves in K^u. Now we discuss the relation between the irreducible $*$-representations of $\mathbb{C}[K^u]_q$ and the symplectic leaves in the Poisson Lie group K^u. Let $\sigma = \sigma_{w,u}$ and $P_Z = \{\lambda \mid \sigma(\lambda, \mu) = 1, \text{ for any } \mu \in P\}$.

Denote by $Z(P, \sigma)$ the center of $A(P, \sigma)$. An irreducible $*$-representation $\chi : A(P, \sigma) \to \operatorname{End} H_\chi$ determines a character $\chi_Z : Z(P, \sigma) \to \mathbb{C}$ of the center, but that character may correspond to several nonequivalent representations of $A(P, \sigma)$. Nevertheless, the following result holds.

PROPOSITION 7.3.1. $\operatorname{Ker} \chi = A(P, \sigma)(\operatorname{Ker} \chi_Z)$.

PROOF. Consider $u \in \operatorname{Ker} \chi$ such that $u \notin A(P, \sigma)(\operatorname{Ker} \chi_Z)$. Let u be of the form $u = \sum_{\lambda \in M} d_\lambda u_\lambda$ with the least possible number of nonzero summands. It is easy to show that there exist $\mu \in P$ and $\lambda_1, \lambda_2 \in M$ such that $\sigma(\lambda_1, \mu) \neq \sigma(\lambda_2, \mu)$ and the element

$$u' = u - \sigma(\lambda_1, \mu)^{-1} u_{-\mu} u u_\mu \in \operatorname{Ker} \chi \setminus A(P, \sigma)(\operatorname{Ker} \chi_Z)$$

is of the form

$$u' = \sum_{\lambda \in M \setminus \lambda_1} d'_\lambda u_\lambda \text{ with } d'_\lambda \neq 0.$$

This contradicts the choice of u. Thus, the proposition is proved. $\square$

Denote $\hat{P} = P/P_Z$. Note that σ uniquely determines another form $\hat{\sigma} : \hat{P} \times \hat{P} \to \mathbb{R}$ and another quantum torus $A(\hat{P}, \hat{\sigma})$.

PROPOSITION 7.3.2. *The unitary equivalence class of irreducible $*$-representations of $A(\hat{P}, \hat{\sigma})$ is unique if and only if $\hat{P}$ is finite.*

PROOF. Since the center of $A(\hat{P}, \hat{\sigma})$ is trivial, the kernel of any irreducible $*$-representation is also trivial. If $\hat{P}$ is finite, then $A(\hat{P}, \hat{\sigma})$ is a finite-dimensional C^*-algebra of type I (cf. [**Dix96**]). In particular, there is a natural bijection between the classes of irreducible $*$-representations and primary $*$-ideals. $\square$

EXERCISE 7.3.3. In the case when $\hat{P}$ is infinite, construct an example of nonequivalent irreducible $*$-representations of $A(\hat{P}, \hat{\sigma})$.

Thus, one can hope that there is a one-to-one correspondence between the symplectic leaves in K^u and the irreducible $*$-representations if and only if $\hat{P}$ is finite. The necessary and sufficient condition for that is given below.

PROPOSITION 7.3.4. $\hat{P}$ is finite if and only if

(7.3.1) $$\log q \theta_{w,u}(\lambda,\mu) \in \pi \mathbb{Q}$$

for any $\lambda, \mu \in P$.

PROOF. It follows from (7.2.4) that $\hat{P}$ is finite if and only if $\sigma(\lambda, \mu) = \sigma_{w,u}(\lambda, \mu)$ is a root of unity for any $\lambda, \mu \in P$. □

We omit the proof of the following theorem, as it will not be used in the book.

THEOREM 7.3.5. (1) If $w \neq 1$, then for almost any u, a symplectic leaf Σ in X_w is equal to X_w provided its rank $m = \frac{1}{2} \dim \Sigma$ is even, and is dense in X_w (but not equal to X_w) if m is odd. Furthermore, for almost any u, we have $P_Z = 0$ and the center of $A(P, \sigma)$ is trivial. For those w and u, all corresponding irreducible $*$-representations of $\mathbb{C}[K^u]_q$ have the same kernel.

(2) If $w = 1$, the corresponding irreducible $*$-representations of $\mathbb{C}[K^u]_q$ are one-dimensional and parameterized by the points of the distinguished maximal torus T.

As we see, for almost any u, there is a natural one-to-one correspondence between the symplectic leaves in K^u and the kernels of the irreducible $*$-representations of $\mathbb{C}[K^u]_q$ if m is even, or between the closures of the symplectic leaves and the kernels of the irreducible $*$-representations of $\mathbb{C}[K^u]_q$ if m is odd.

7.4. Representations of $\mathbb{C}[K(0,u)]_q$. Now we consider the quantized algebra of functions $\mathbb{C}[K(0,u)]_q$. Let $L_{(\lambda,\mu)}$ be the span of the matrix elements of the form $c^{\Lambda}_{-\lambda,i,\mu,j}$. Clearly, the decomposition

$$\mathbb{C}[K(0,u)]_q = \bigoplus_{(\lambda,\mu) \in P \times P} L_{(\lambda,\mu)}$$

defines a structure of $P \times P$-graded algebra in $\mathbb{C}[K(0,u)]_q$. Denote by $\psi = \psi_q$ the set of the maximal two-sided $P \times P$-graded $*$-ideals in $\mathbb{C}[K(0,u)]_q$.

REMARK 7.4.1. By definition, $\mathbb{C}[K(0,u)]_q$ and $\mathbb{C}[K] = \mathbb{C}[K(0,0)]_1$ coincide as sets. It follows from the definition of the algebra structure in $\mathbb{C}[K(0,u)]_q$ that $\psi_q = \psi_1$ with respect to this identification.

DEFINITION 7.4.2. For any irreducible $*$-representation π of $\mathbb{C}[K(0,u)]_q$, we write $\alpha = \mathrm{Gr}(\pi)$ if $\alpha \in \psi$ is the maximal element in the set $\{\beta \in \psi \mid \beta \subset \mathrm{Ker}\,\pi\}$.

Consider the following sets:

$$\Gamma_\alpha = \{(\lambda,\mu) \in P \times P \mid L_{(\lambda,\mu)} \not\subset \alpha\},$$
$$\Gamma(\pi) = \{(\lambda,\mu) \in P \times P \mid L_{(\lambda,\mu)} \not\subset \mathrm{Ker}\,\pi\}.$$

PROPOSITION 7.4.3. (1) $\Gamma(\pi)$ is a sublattice in $P \times P$.
(2) $\Gamma(\pi) = \Gamma(\alpha)$, where $\alpha = \mathrm{Gr}(\pi)$.

PROOF. (1) It is clear that if $f_1, f_2 \in L_{(\lambda,\mu)}$, then the element $f_1^* \circ f_2 = f_2 \circ f_1^*$ is central. Choose $f \in L_{(\lambda,\mu)}$ and $g \in L_{(\gamma,\delta)}$ such that $\pi(f) \neq 0$ and $\pi(g) \neq 0$. Then $f \circ g \in L_{(\lambda+\gamma, \mu+\delta)}$ and

$$\pi(f \circ g)\pi(g^* \circ f^*) = \pi(f)\pi(g \circ g^*)\pi(f) = \pi(f \circ f^*)\pi(g \circ g^*) \neq 0.$$

Hence, we get that $(\lambda_\gamma, \mu + \delta) \in \Gamma(\pi)$.

(2) Since $\alpha \in \mathrm{Ker}\,\pi$, we have $\Gamma(\pi) \subset \Gamma_\alpha$. To prove the inclusion $\Gamma_\alpha \subset \Gamma(\pi)$, assume that there exists $(\lambda, \mu) \in \Gamma_\alpha$ such that $(\lambda, \mu) \notin \Gamma(\pi)$. Choose $f \in L_{(\lambda,\mu)} \setminus \alpha$

such that $\pi(f) = 0$, and consider the minimal ideal $\alpha_1 \in \psi$ which contains both α and f. Clearly, we have that $\alpha \subsetneq \alpha_1 \subset \operatorname{Ker} \pi$, but this contradicts the fact that α is maximal. $\square$

PROPOSITION 7.4.4. *If π is an irreducible $*$-representation of $\mathbb{C}[K(0,u)]_q$ in a separable Hilbert space H, then $\dim \pi(L_{(\lambda,\mu)}) = 1$ for any $(\lambda,\mu) \in \Gamma(\pi)$.*

PROOF. By definition of $\Gamma(\pi)$, we have $\pi(L_{(\lambda,\mu)}) \neq \{0\}$. Choose $f_1, f_2 \in L_{(\lambda,\mu)}$ such that $\pi(f_1) \neq 0$ and $\pi(f_2) \neq 0$. Since both $f_1 \circ f_2^*$ and $f_2^* \circ f_2$ are central, we have
$$\pi(f_1) = \pi(f)\pi(f_2^*)\,\pi(f_2)\pi(f_2^* \circ f_2)^{-1} = d\pi(f_2),$$
where $d = \pi(f_2^* \circ f_2)^{-1}\pi(f_1 \circ f_2^*) \in \mathbb{C}$. $\square$

PROPOSITION 7.4.5. *For any $(\lambda,\mu) \in \Gamma_\alpha = \Gamma(\pi)$, there exists $a_{\lambda,\mu} \in \pi(L_{(\lambda,\mu)})$ such that the following relations hold for any $(\lambda,\mu), (\lambda',\mu') \in \Gamma_\alpha$:*

(7.4.1) $$a_{\lambda,\mu}^* = a_{-\lambda,-\mu} = a_{\lambda,\mu}^{-1},$$
(7.4.2) $$a_{\lambda,\mu}a_{\lambda',\mu'} = \sigma((\lambda,\mu),(\lambda',\mu'))a_{\lambda+\lambda',\mu+\mu'}.$$

PROOF. For a set of generators (λ,μ) of the $\mathbb{Z}$-module Γ_α, choose $a_{\lambda,\mu}$ so that they satisfy (7.4.1) (clearly, this can be done). For any other $(\lambda,\mu) \in \Gamma_\alpha$, define $a_{\lambda,\mu}$ by (7.4.2). $\square$

Define $\sigma_{\alpha,u} : \Gamma_\alpha \times \Gamma_\alpha \to \mathbb{C}$ by
$$\sigma_{\alpha,u}((\lambda,\mu),(\lambda',\mu')) = q^{i((\mu,\mu')_u - (\lambda,\lambda')_u)},$$
and consider the quantum torus $A(\Gamma_\alpha, \sigma_{\alpha,u})$. Choose $a_{\lambda,\mu} \in \pi(L_{(\lambda,\mu)})$ for any $(\lambda,\mu) \in \Gamma_\alpha$ as in Proposition 7.4.5. For any $f \in L_{(\lambda,\mu)}$ and $U_{\lambda,\mu}$ define
$$\pi_0(f) = \pi(f)a_{\lambda,\mu}^{-1} \in \mathbb{C}, \quad \chi(U_{\lambda,\mu}) = a_{\lambda,\mu}.$$
Extend π_0 and χ linearly to the maps $\pi_0 : \mathbb{C}[K] \to \mathbb{C}$ and $\chi : A(\Gamma_\alpha, \sigma_{\alpha,u}) \to \operatorname{End} H$ respectively (see Remark 7.4.1).

THEOREM 7.4.6. *The maps π_0 and χ are irreducible $*$-representations of $\mathbb{C}[K]$ and $A(\Gamma_\alpha, \sigma_{\alpha,u})$ respectively.*

PROOF. It follows from (7.4.1), (7.4.2), and the definition of the multiplication in $\mathbb{C}[K(0,u)]_q$ that π_0 and χ are $*$-representations. Clearly, π_0 is irreducible since it is one-dimensional. Finally, χ is irreducible, since π is. $\square$

THEOREM 7.4.7. *Let $\pi_0 : \mathbb{C}[K] \to \mathbb{C}$ and $\chi : A(\Gamma_\alpha, \sigma_{\alpha,u}) \to \operatorname{End} H$ be irreducible $*$-representations, where $\alpha = \operatorname{Gr}(\pi)$. Then the map $\pi : \mathbb{C}[K(0,u)]_q \to \operatorname{End} H$ defined on $f \in L_{(\lambda,\mu)}$ by*
$$\pi(f) = \begin{cases} \pi_0(f)\chi(U_{\lambda,\mu}) & \text{if } f \in L_{(\lambda,\mu)}, (\lambda,\mu) \in \Gamma_\alpha, \\ 0 & \text{if } f \in L_{(\lambda,\mu)}, (\lambda,\mu) \notin \Gamma_\alpha, \end{cases}$$
and extended linearly to $\mathbb{C}[K(0,u)]_q$ is an irreducible $*$-representation.

PROOF. By definition of the multiplication in $\mathbb{C}[K(0,u)]_q$ and $A(\Gamma_\alpha, \sigma_{\alpha,u})$, π is a $*$-representation. The irreducibility is obvious. $\square$

DEFINITION 7.4.8. Pairs (π_0, χ) and (π_0', χ') (as in Theorem 7.4.7) are said to be *equivalent* if $\mathrm{Gr}(\pi_0) = \mathrm{Gr}(\pi_0')$ and there exists a linear real-valued function φ on Γ_α such that

$$\pi_0'(f) = e^{-i\varphi(\lambda,\mu)}\pi_0(f), \quad \chi'(U_{\lambda,\mu}) = e^{i\varphi(\lambda,\mu)}\chi(U_{\lambda,\mu})$$

for any $(\lambda, \mu) \in \Gamma_\alpha$ and $f \in L_{(\lambda,\mu)}$.

The following statement is obvious.

PROPOSITION 7.4.9. *Pairs (π_0, χ) and (π_0', χ') determine the same irreducible $*$-representation π of $\mathbb{C}[K(0,u)]_q$ if and only if they are equivalent.*

Thus, the irreducible $*$-representations of $\mathbb{C}[K(0,u)]_q$ are parameterized by the pairs (π_0, χ) up to the equivalence described in Definition 7.4.8. Here π_0 is an irreducible $*$-representation of $\mathbb{C}[K]$ (that is, a point of K), and χ is an irreducible $*$-representation of the quantum torus $A(\Gamma_\alpha, \sigma_{\alpha,u})$, where $\alpha = \mathrm{Gr}(\pi_0)$.

8. Representations of formal quantized algebras of functions

The formal quantized algebras of functions $\mathbb{C}[G]_h$ and $\mathbb{C}[K]_h$ were defined in Remark 1.2.5. One can develop representation theory of these algebras which is parallel to the representation theory for $\mathbb{C}[G]_q$ and $\mathbb{C}[K]_q$. We present it as a series of exercises.

EXERCISE 8.0.1. (1) Let $\mathcal{O}_0 \subset \mathbb{C}[[h]]$ be the subalgebra of the Taylor series of germs at $h - 0$ of complex analytic functions. Show that the family of Hopf algebras $\mathbb{C}[G]_q$, where $0 < |q| < 1, q = e^{\frac{h}{2}}$, yields a Hopf algebra over $\mathcal{O}_0$ which we denote by $\mathbb{C}[G]_{q,\mathcal{O}_0}$.

(2) Let $\mathcal{O}_0^{\mathbb{R}}$ be the algebra of germs at $h = 0$ of real analytic functions on $\mathbb{R}$. Taking the restrictions of the elements of $\mathcal{O}_0$ to $\mathbb{R}$, we define the Hopf algebra $\mathbb{C}[G]_{q,\mathcal{O}_0^{\mathbb{R}}}$. Prove that both $\mathbb{C}[G]_{q,\mathcal{O}_0}$ and $\mathbb{C}[G]_{q,\mathcal{O}_0^{\mathbb{R}}}$ have a Hopf $*$-algebra structure such that $h^* = \bar{h}$ for $\mathbb{C}[G]_{q,\mathcal{O}_0}$ and $h^* = h$ for $\mathbb{C}[G]_{q,\mathcal{O}_0^{\mathbb{R}}}$.

EXERCISE 8.0.2. Define $\mathbb{C}[K]_{q,\mathcal{O}_0^{\mathbb{R}}}$ as a pair $(\mathbb{C}[G]_{q,\mathcal{O}_0^{\mathbb{R}}}, *)$ (see Exercise 8.0.1). Show that

$$\begin{aligned}\mathbb{C}[G]_h &\simeq \mathbb{C}[[h]] \otimes_{\mathbb{C}} \mathbb{C}[G]_{q,\mathcal{O}_0}, \\ \mathbb{C}[K]_h &\simeq \mathbb{C}[[h]] \otimes_{\mathbb{C}} \mathbb{C}[K]_{q,\mathcal{O}_0^{\mathbb{R}}}.\end{aligned}$$

Let A be a commutative ring, and B an A-algebra. We say that an A-module V is a B-module if it is free as an A-module and there is an action of B on V compatible with the structure of A-module.

If A is an algebra with involution $a \mapsto \bar{a}$, we say that a bilinear A-form $\langle , \rangle : V \times V \to A$ on an A-module V is Hermitian if

$$\langle av_1, v_2 \rangle = \langle v_1, \bar{a}v_2 \rangle = a\langle v_1, v_2 \rangle, \quad \langle v_2, v_1 \rangle = \overline{\langle v_1, v_2 \rangle},$$

for any $a \in A$ and $v_1, v_2 \in V$. We say that a bilinear A-form on V is nondegenerate if there is no nonzero element $v \in V$ such that $\langle v, w \rangle = 0$ for any $w \in V$.

If A is a complete topological algebra, we understand everything in the topological setup. That is, B is a complete topological A-algebra, and V is a B-module which is a comlete topologically free A-module.

In both the algebraic and topological case we say that an A-module is Hermitian if it is equipped with a nondegenerate Hermitian form. If a Hermitian A-module

V is a B-module and B is an A-algebra with an involution $b \mapsto b^*$ (so that, in particular, $(ab)^* = \bar{a}b^*$ for any $a \in A, b \in B$), we say that V is a Hermitian B-module provided that
$$\langle bv_1, v_2 \rangle = \langle v_1, b^* v_2 \rangle$$
for any $b \in B$ and $v_1, v_2 \in V$.

DEFINITION 8.0.3. (1) We say that an Hermitian $\mathbb{C}[G]_{q, \mathcal{O}_0^{\mathbb{R}}}$-module V is *unitarizable* if for any nonzero $v \in V$, $\langle v, v \rangle \in \mathcal{O}_0^{\mathbb{R}}$ is a positive real analytic function on some interval $-1 < -r_0 \le h \le r_0 < 1$.

(2) We treat $A = \mathbb{C}[[h]]$ as an algebra with involution $\overline{\lambda h} = \bar{\lambda} h$, where $\lambda \in \mathbb{C}$, and $B = \mathbb{C}[G]_h$ as an A-algebra. We say that a $\mathbb{C}[G]_h$-module V is *unitarizable* if there exists a unitarizable $\mathbb{C}[G]_{q, \mathcal{O}_0^{\mathbb{R}}}$-module V_0 such that

(8.0.1) $$V \simeq \mathbb{C}[[h]] \otimes_{\mathbb{C}} V_0.$$

(3) For any Hermitian $\mathcal{O}_0^{\mathbb{R}}$-module, its completion with respect to the natural topology is called *Hilbert $\mathcal{O}_0^{\mathbb{R}}$-space*.

(4) We say that a unitarizable $\mathbb{C}[G]_h$-module V gives rise to a *$*$-representation* of $\mathbb{C}[K]_h$ if there exists a Hilbert $\mathcal{O}_0^{\mathbb{R}}$-space V_0 such that (8.0.1) holds.

EXERCISE 8.0.4. Prove that all the results of Section 5 about the $\mathbb{C}[G]_q$-modules and $\mathbb{C}[K]_q$-modules remain true for $\mathbb{C}[G]_h$-modules and $\mathbb{C}[K]_h$-modules respectively.

REMARK 8.0.5. One can define the notion of a primitive ideal in both $\mathbb{C}[G]_h$ and $\mathbb{C}[K]_h$. Then all the results of Section 5 formulated for primitive ideals of $\mathbb{C}[G]_q$ and $\mathbb{C}[K]_q$ still hold. In fact, the reason why we have used algebras over $\mathcal{O}_0^{\mathbb{R}}$ is that now the proofs from Section 5 can be repeated almost word for word.

9. Historical remarks

Representation theory of the quantized algebra of regular functions on the quantum group $SU(2)$ (cf. Sections 1–4) was first developed in [**SV86**] (a later, abridged version can be found in [**SV88**]). In particular, Theorem 3.0.1 was proved there. The theory of unitarizable $\mathbb{C}[G]_q$-modules and the equivalent theory of $*$-representations of $\mathbb{C}[K]_q$ presented in Section 5 are based on the original paper by one of the authors [**Soi90**]. This is a special case of the correspondence between the primitive ideals in $\mathbb{C}[G]_q$ and the symplectic leaves of the standard Poisson Lie group structure on G. The general case was done in [**Jos95**] and [**HL93**] along the lines of [**Soi90**]. Slight improvements of the original proofs of [**Soi90**] offered in those papers were used in Section 5. The case of the twisted quantized algebras of functions (cf. Section 7) was considered in [**Lev91**] and [**LS91b**].

Algebras of functions on quantum groups were discussed in several books and papers, although from different points of view. We can refer the interested reader to Chapter 29 of [**Lus93**], to the book [**Jos95**], or to the paper [**Wo87a**]. In [**Wo87a**], the axioms for compact matrix quantum groups were suggested. We should mention that the algebra of functions on the quantized group $SU(2)$ was introduced independently in [**Wo87b**]. Another approach to the topic can be found in [**FRT88**]. Roughly speaking, the main difference between these two approaches can be formulated as the following question: should one start with the quantized universal enveloping algebra and then recover the algebra of functions on quantum group via representation theory, or should one start with the quantized algebra of functions described similarly to our Section 3? The latter approach can be developed successfully if one has enough information about R-matrices of the corresponding quantum groups. This "R-matrix approach" is closer to the original works of physicists, but it requires many formal computations with the "R-matrix algebra." We use the alternative approach and start with the quantized universal enveloping algebra. In turn, it requires some knowledge of the structure of representations (see, for example, Theorem 3.0.4). According to Chapter 2, this structure is similar to the case $q = 1$, which is very well known.

CHAPTER 4

Quantum Weyl Group and the Universal Quantum R-Matrix

In this chapter we define quantized elements of the Weyl group as Gelfand–Naimark–Segal states on the quantized algebra of functions. We start with describing the quasi-classical picture in Section 1, and then use the results of Chapter 3 to quantize it. The quantum Weyl group is defined then as a certain extension of the quantized universal enveloping algebra. We use it to derive an explicit formula for the universal quantum R-matrix.

1. Motivations: Weyl group in the quasi-classical picture

Let G be a connected and simply connected simple complex Lie group, $\mathfrak{g} = \mathrm{Lie}\, G$ its Lie algebra, $K \subset G$ a maximal compact subgroup, and $T \subset K$ a maximal torus (see Chapter 1, Section 5.1). We keep the same conventions and use the same notation as in Chapter 2.

We recall that one can define the Weyl group W in two completely different ways. We will intentionally denote the Weyl groups resulting from the following two constructions by different symbols, namely, W_1 and W_2.

DEFINITION 1.0.1. The *Weyl group* W_1 is the quotient group $N(T)/T$, where $N(T)$ is the normalizer in K of the maximal torus T.

DEFINITION 1.0.2. The *Weyl group* W_2 is the subgroup of $\mathrm{Aut}\,(\mathfrak{h}^*)$ generated by the simple reflections

(1.0.1) $$s_i : \lambda \mapsto \lambda - \frac{2(\lambda, \alpha_i)}{(\alpha_i, \alpha_i)} \alpha_i, \quad i = 1, 2, \ldots, n.$$

The following classical result is well known (cf. [**Bou89**, Chapters 4–6]).

PROPOSITION 1.0.3. *The groups W_1, W_2 are both isomorphic to the group W generated by $s_1, s_2, \ldots, s_n$, subject to the relations*

(1.0.2) $$s_i^2 = 1, \quad i = 1, 2, \ldots, n,$$

(1.0.3) $$s_i s_j s_i \cdots = s_j s_i s_j \cdots, \quad i \neq j,$$

with both sides of (1.0.3) containing m_{ij} factors, where m_{ij} is equal to $2, 3, 4, 6$ if $a_{ij}a_{ji}$ is equal to $0, 1, 2, 3$ respectively.

It is also well known (cf. [**Bou89**, Chapters 4–6]) that one can choose a subgroup $\bar{W}_1 \subset N(T)$ such that $W_1 = p(\bar{W}_1)$, where p is the natural projection $N(T) \to N(T)/T$.

Consider the group algebra $\mathbb{C}\bar{W}_1$. One can say that it is generated by the delta functions δ_w supported at $w \in \bar{W}_1$. Thinking of delta-functions as linear functionals on $\mathbb{C}[G]$ (or on $\mathbb{C}[K]$), we get a Hopf subalgebra in the full dual Hopf algebra $\mathbb{C}[G]^*$

generated by the elements of $\bar{W}_1$. Of course, the property $\Delta(w) = w \otimes w$ is satisfied for any $w \in \bar{W}_1$. As is well known (cf. [**Ste72**]), one can choose representatives for s_i ($i = 1, 2, \ldots, n$) in $\bar{W}_1$ such that they satisfy (1.0.3), but not (1.0.2).

Let us fix $w \in \bar{W}_1$ and consider the automorphism r_w of $\mathbb{C}[G]$ given by $r_w : f(g) \mapsto f(gw)$. Suppose that we fixed the standard Poisson Lie group structure on G given by the classical r-matrix r from Chapter 1, (4.1.7). Note that it is compatible with the embedding $K \subset G$, so that we get a Poisson Lie group structure on K, too.

Since the map $g \mapsto gw$ is not an automorphism of the Poisson Lie group G, the map r_w is not an automorphism of the Poisson Hopf algebra $\mathbb{C}[G]$. Nevertheless, one can introduce a new Poisson structure on G such that $g \mapsto gw$ becomes a Poisson map (but not a morphism of Poisson Lie groups). This can be made very transparent in the case when $w = w_0$ is the longest element of the Weyl group. Then the original Poisson bracket

$$\{f_1, f_2\} = \sum_{\alpha,\beta} r_{\alpha\beta} \left(\partial_\alpha f_1 \partial_\beta f_2 - \partial'_\alpha f_1 \partial'_\beta f_2 \right)$$

(cf. Chapter 1, Remark 4.1.4) becomes

$$\{f_1, f_2\}_s = \sum_{\alpha,\beta} r_{\alpha\beta} \left(\partial_\alpha f_1 \partial_\beta f_2 + \partial'_\alpha f_1 \partial'_\beta f_2 \right).$$

Here ∂_α (resp. ∂'_α) are the right (resp. left) invariant vector fields on G corresponding to the choice of a basis $\{I_\alpha\}$ in $\mathfrak{g}$.

REMARK 1.0.4. As we observe, Definition 1.0.1 of the Weyl group yields a group $\bar{W}_1$ related to the Poisson Lie group structure on G. On the other hand, Definition 1.0.2 is purely algebraic. We will see later that there is a remarkable possibility to quantize $\bar{W}_1$ in such a way that we will get a Hopf algebra called the *quantum Weyl group*. However, there is no canonical way to quantize W_2 to get a Hopf algebra.

Sometimes, people keep the relation (1.0.3) untouched, deforming (1.0.2) instead. For example, if one replaces (1.0.2) with

$$\left(s_i - q_i^2 \right) (s_i + 1) = 0, \quad i = 1, 2, \ldots, n,$$

one obtains the Hecke algebra of $\mathfrak{g}$ (cf. [**Lus93**]). We will see later that the quantum Weyl group is closely related to this deformation of the group algebra $\mathbb{C}W_2$.

As further motivation, we discuss a relation between the Weyl group W_1 and the classical r-matrix; this relation gives rise to a relation between the quantum Weyl group and the universal quantum R-matrix. Let

$$\varphi(a) = [a \otimes 1 + 1 \otimes a, r], \quad a \in \mathfrak{g},$$

be the 1-cocycle on $\mathfrak{g}$ with coefficients in $\mathfrak{g} \otimes \mathfrak{g}$ corresponding to the classical r-matrix r (see (2.3.3) in Chapter 1). When we replace the Poisson structure on G to make $g \mapsto gw$ a Poisson map, the corresponding cocycle also changes. In particular, when $w = w_0$ is the longest element of the Weyl group, the new cocycle is equal to

(1.0.4) $$\varphi_{w_0}(a) = \varphi(a) + 2(a \otimes 1 + 1 \otimes a)r.$$

The "quantum" version of this cocycle is a new comultiplication in $U_h\mathfrak{g}$ given by

(1.0.5) $$\Delta_{w_0}(a) = \Delta(a)R,$$

where Δ is the standard comultiplication (cf. Chapter 2, (1.2.4)), and R is the universal quantum R-matrix. Indeed, it is easy to see that

(1.0.6) $$\Delta_{w_0}(a) - \Delta'_{w_0}(a) = h\varphi_{w_0}(a) + O\left(h^2\right),$$

where Δ'_{w_0} is the opposite comultiplication. Note that, even though Δ_{w_0} does not define a Hopf algebra structure on $U_h\mathfrak{g}$, it is coassociative due to formulas (2.1.4)–(2.1.5) in Chapter 2.

If we accept the above motivations (to be made rigorous later), then we see that a quantum analog of the automorphism $g \mapsto gw_0$ should be a coalgebra automorphism $(U_h\mathfrak{g}, \Delta) \mapsto (U_h\mathfrak{g}, \Delta_{w_0})$ such that

(1.0.7) $$a \mapsto a\bar{w}'_0$$

for some element $\bar{w}'_0$ in $U_h\mathfrak{g}$ (or in some completion of $U_h\mathfrak{g}$). Then

(1.0.8) $$\Delta(a)\Delta\left(\bar{w}'_0\right) R = \Delta(a)\left(\bar{w}'_0 \otimes \bar{w}'_0\right).$$

Therefore,

(1.0.9) $$\Delta\left(\bar{w}'_0\right) = R^{-1}\left(\bar{w}'_0 \otimes \bar{w}'_0\right).$$

A similar hypothetical formula can be obtained for a quantum analog of any element of $\bar{W}_1$. In particular, we may expect the following formula to hold for a quantum analog $\bar{s}'_i$ of a simple reflection s_i:

(1.0.10) $$\Delta\left(\bar{s}'_i\right) = R_i^{-1}\left(\bar{s}'_i \otimes \bar{s}'_i\right),$$

where R_i is the universal quantum R-matrix of the Hopf subalgebra $U_h\left(\mathfrak{sl}(2,\mathbb{C})_i\right)$ in $U_h\mathfrak{g}$ generated by X_i^+, X_i^-, H_i (see Chapter 2, Proposition 2.3.12, for the explicit formula).

REMARK 1.0.5. Note that yet another motivation for (1.0.9) can be derived from the formula

(1.0.11) $$(\mathrm{Ad}_{w_0} \otimes \mathrm{Ad}_{w_0})\varphi(a) = -\varphi\left(\mathrm{Ad}_{w_0}(a)\right),$$

valid for any $a \in \mathfrak{g}$. Here $\mathrm{Ad}_{w_0}(a) = w_0 a w_0^{-1}$. Thus, $\mathrm{Ad}_{\bar{w}'_0}$ must be a coalgebra antiautomorphism of $U_h\mathfrak{g}$. If we now assume that (1.0.9) holds, the formula (2.1.1) from Chapter 2 can be obtained (which we know to hold already).

Finally, we remark that, if proved, the formula (1.0.9) could allow us to derive an explicit formula for the universal quantum R-matrix R. As the reader will see, we choose a more direct way to derive it, namely, we use the approach based on the quantum double construction from Chapter 2. Still, the quantum Weyl group will play an essential role in our computations.

2. Quantum Weyl group: definitions

Let V_{s_i} be the unitarizable $\mathbb{C}[G]_q$-module defined in Chapter 3, Section 6.1. It contains a highest vector e_0, which generates a unitarizable $\mathbb{C}[SL_2(\mathbb{C})]_{q_i}$-module $V_\mathcal{D}$. Let w be an element of the Weyl group W of $\mathfrak{g}$, and $w = s_{i_1} s_{i_2} \cdots s_{i_k}$ a reduced decomposition of w. By Theorem 6.2.1 of Chapter 3, the $\mathbb{C}[G]_q$-module $V_{s_{i_1}} \otimes V_{s_{i_2}} \otimes \cdots \otimes V_{s_{i_k}}$ is unitarizable and simple, generated by the highest vector $e_w = e_0^{\otimes l(w)}$. As in Chapter 3, we denote by π_{s_i} the irreducible $*$-representation of $\mathbb{C}[G]_q$ corresponding to V_{s_i}.

DEFINITION 2.0.1. For any $w \in W$, the *quantized element* $\bar{w}$ is the linear functional on $\mathbb{C}[G]_q$ defined by

(2.0.1) $$\bar{w}(f) = \langle (\pi_{s_{i_1}} \otimes \pi_{s_{i_2}} \otimes \cdots \otimes \pi_{s_{i_k}})(f) e_w, e_w \rangle,$$

where $f \in \mathbb{C}[G]_q$ and $\langle\,,\,\rangle$ is the scalar product in the unitarizable $\mathbb{C}[G]_q$ module $V_{s_{i_1}} \otimes \cdots \otimes V_{s_{i_k}}$.

One readily recognizes in (2.0.1) the so-called Gelfand–Naimark–Segal state corresponding to the representation $\pi_{s_{i_1}} \otimes \cdots \otimes \pi_{s_{i_k}}$ and the vector e_w in the space of the representation.

PROPOSITION 2.0.2. *The linear functional $\bar{w}$ does not depend on the choice of a reduced decomposition of $w \in W$. Moreover, one has that $\bar{w} = \bar{s}_{i_1} \bar{s}_{i_2} \cdots \bar{s}_{i_k}$ (the product in the Hopf algebra dual to $\mathbb{C}[G]_q$).*

PROOF. As for the elements of the classical Weyl group, it suffices to prove that

(2.0.2) $$\bar{s}_i \bar{s}_j \bar{s}_i \cdots = \bar{s}_j \bar{s}_i \bar{s}_j \cdots$$

for any $i \neq j$. Here $\bar{s}_k$ stands for a quantized simple reflection s_k, and the products on both sides of (2.0.2) consist of m_{ij} factors, where m_{ij} is the order of $s_i s_j$ in W. But this follows from Proposition 6.2.2 of Chapter 3 if we note that the isomorphism defined there can be chosen to preserve e_w. □

2.1. The case of $\mathbb{C}[SL_2(\mathbb{C})]_{q_i}$. Let V_m be the $m+1$-dimensional unitarizable simple $U_{q_i}\mathfrak{sl}(2,\mathbb{C})$-module given by (2.3.5)–(2.3.7) in Chapter 3. Here we have $q_i = q^{\frac{(\alpha_i,\alpha_i)}{2}}$ and $m = 0, 1, 2, \ldots$. Denote by t^l_{jk} the linear basis of matrix elements in $\mathbb{C}[SL_2(\mathbb{C})]_{q_i}$ corrresponding to the orthonormal basis $\{e_k\}$ in V_m (cf. Chapter 3, (2.3.5)–(2.3.7)). Denote by $\bar{s}_0$ the only nontrivial quantized simple reflection.

PROPOSITION 2.1.1. *The quantized simple reflection $\bar{s}_0$ satisfies the following system of equations:*

(2.1.1) $\quad\quad\quad \bar{s}_0(f t_{11}) = \bar{s}_0(t_{22} f) = 0,$

(2.1.2) $\quad\quad\quad \bar{s}_0(t_{12} f) = q_i \bar{s}_0(f),$

(2.1.3) $\quad\quad\quad \bar{s}_0(t_{21} f) = -\bar{s}_0(f),$

for any $f \in \mathbb{C}[SL_2(\mathbb{C})]_{q_i}$ (the matrix elements $t_{ij} = t^1_{ij}$ were defined in Theorem 3.0.1 of Chapter 3).

PROOF. The assertion follows immediately from the definition of the reflection $\bar{s}_0$ and formulas (4.1.3)–(4.1.7) of Chapter 3 with $\varphi = \pi$. □

COROLLARY 2.1.2 (Exercise). *The system of equations (2.1.1)–(2.1.3) has a unique solution $\bar{s}_0$ such that $\bar{s}_0(1) = 1$. It is given by the following explicit formula:*

(2.1.4) $$\bar{s}_0\left(t^m_{jk}\right) = \delta_{j,m+1-j}(-1)^{m+1-j} q_i^{m(m+1)-j(j+1)}.$$

Consider the elements $E_i = X_i^+ K_i^{-1}$ and $F_i = X_i^- K_i$ of $U_{q_i}\mathfrak{sl}(2,\mathbb{C})$. They can be regarded as linear functionals on $\mathbb{C}[SL_2(\mathbb{C})]_{q_i}$. Moreover, if $X_i^+, X_i^-, K_i^{\pm}$ form an $U_{q_i}\mathfrak{sl}(2,\mathbb{C})$-triple in $U_q\mathfrak{g}$, the elements E_i and F_i can be regarded as linear functionals on $\mathbb{C}[G]_q$.

2. QUANTUM WEYL GROUP: DEFINITIONS

PROPOSITION 2.1.3. *The following identities hold in* $(\mathbb{C}[SL_2(\mathbb{C})]_{q_i})^*$, *the Hopf algebra dual to* $\mathbb{C}[SL_2(\mathbb{C})]_{q_i}$:

(2.1.5) $\quad\quad\quad\quad\quad\quad\quad \bar{s}_i E_i \bar{s}_i^{-1} = -K_i^{-1} F_i,$

(2.1.6) $\quad\quad\quad\quad\quad\quad\quad \bar{s}_i F_i \bar{s}_i^{-1} = -E_i K_i,$

(2.1.7) $\quad\quad\quad\quad\quad\quad\quad \bar{s}_i K_i \bar{s}_i^{-1} = K_i^{-1}.$

PROOF. The assertion follows from (2.1.4) and formulas (2.3.5)–(2.3.7) in Chapter 3. $\square$

For any admissible $U_{q_i}\mathfrak{sl}(2,\mathbb{C})$-module V and a pair $(l,v) \in V^* \oplus V$, we have a matrix element $c_{l,v} \in \mathbb{C}[SL_2(\mathbb{C})]_{q_i}$. Recall that such matrix elements generate the quantized algebra of functions. For any linear operator $A : V \to V$, we define a linear functional $\eta_A \in (\mathbb{C}[SL_2(\mathbb{C})]_{q_i})^*$ by

(2.1.8) $\quad\quad\quad\quad\quad\quad\quad \eta_A(c_{l,v}) = l(Av).$

Let $V = \bigoplus_\lambda V_\lambda$ be the weight decomposition of an admissible $U_{q_i}\mathfrak{sl}(2,\mathbb{C})$-module V. We define $t_i \in \mathrm{End}_\mathbb{C} V$ such that

(2.1.9) $\quad\quad\quad\quad\quad\quad\quad t_i(v_\lambda) = q^{-\frac{\lambda(H_i)^2}{2(\alpha_i,\alpha_i)}} v_\lambda,$

for any $\lambda \in \mathfrak{h}^*$ and $v_\lambda \in V_\lambda$. It defines a compatible family of linear functionals on $\mathbb{C}[SL_2(\mathbb{C})]_{q_i}$ (one for each $U_{q_i}\mathfrak{sl}(2,\mathbb{C})$-module V). Thus we get a linear functional on $\mathbb{C}[SL_2(\mathbb{C})]_{q_i}$ which we also denote by t_i.

Similarly, for any admissible $U_{q_i}\mathfrak{sl}(2,\mathbb{C})$-modules V_1 and V_2, we define $D_i \in \mathrm{End}_\mathbb{C}(V_1 \otimes V_2)$ such that

(2.1.10) $\quad\quad\quad\quad\quad\quad D_i(x_\lambda \otimes y_\mu) = q_i^{\frac{\lambda(H_i)\mu(H_i)}{(\alpha_i,\alpha_i)}} x_\lambda \otimes y_\mu.$

Then D_i also defines a linear functional on $(\mathbb{C}[SL_2(\mathbb{C})]_{q_i})^{\otimes 2}$, which we denote also by D_i.

Consider the expression

(2.1.11) $\quad\quad\quad\quad\quad\quad R_i = \left(\sum_{n=0}^{\infty} \frac{(1-q_i^{-2})^n}{(n)_{q_i^{-2}}!} E_i^n \otimes F_i^n\right) D_i,$

where $(n)_q = \frac{1-q^n}{1-q}$ and $(n)_q! = (1)_q (2)_q \cdots (n)_q$. Since both E_i and F_i act by locally nilpotent linear operators in any admissible $U_{q_i}\mathfrak{sl}(2,\mathbb{C})$-module, we conclude that R_i is well defined as a linear functional on $(\mathbb{C}[SL_2(\mathbb{C})]_{q_i})^{\otimes 2}$. Moreover, it is in fact an element of $(\mathbb{C}[SL_2(\mathbb{C})]^*_{q_i})^{\otimes 2}$, where $*$ means the entire dual linear space. The reader can compare (2.1.11) with Chapter 2, (2.3.21).

Finally, define $\bar{s}'_i \in (\mathbb{C}[SL_2(\mathbb{C})]_{q_i})^*$ by

(2.1.12) $\quad\quad\quad\quad\quad\quad\quad \bar{s}'_i = t_i \bar{s}_i.$

PROPOSITION 2.1.4. *The following identities hold:*

(2.1.13) $\quad\quad\quad\quad \Delta(\bar{s}'_i) = R_i^{-1}(\bar{s}'_i \otimes \bar{s}'_i),$

(2.1.14) $\quad\quad\quad\quad \varepsilon(\bar{s}'_i) = \varepsilon(\bar{s}_i) = 1,$

(2.1.15) $\quad\quad\quad\quad S(\bar{s}'_i) = (\bar{s}'_i)^{-1} \sum_{n=0}^{\infty} \frac{(1-q_i^{-2})^n}{(n)_{q_i^{-2}}!} E_i^n t_i S(F_i^n).$

Here Δ, ε, S are respectively the comultiplication, counit, antipode in the Hopf algebra $(\mathbb{C}[SL_2(\mathbb{C})]_{q_i})^*$, and the right-hand side of (2.1.15) is understood as an element in $(\mathbb{C}[SL_2(\mathbb{C})]_{q_i})^*$.

PROOF. The proof is a straightforward computation based on (2.1.4) and Chapter 3, (2.3.5)–(2.3.7). We leave the details to the reader. □

EXERCISE 2.1.5. Show that $(\bar{s}'_i)^2$ coincides, as an element in $(\mathbb{C}[SL_2(\mathbb{C})]_{q_i})^*$, with the linear functional defined by $\Gamma_V u_V : V \to V$ for any admissible $U_{q_i}\mathfrak{sl}(2,\mathbb{C})$-module V (see Chapter 2, Lemma 7.3.2). In particular, it follows that the element

$$(\bar{s}'_i)^2 = q_i^{-2\rho} u$$

is central in $(\mathbb{C}[SL_2(\mathbb{C})]_{q_i})^*$.

2.2. General case. Recall from Chapter 2, Section 3.1 that a Hopf algebra can be equipped canonically with an $A \otimes A$-module structure such that

(2.2.1) $$(a \otimes b) \cdot x = axS(b),$$

where $a, b, x \in A$, and $S : A \to A$ is the antipode. The adjoint action of A on itself is defined by (cf. Definition 1.2.3 of Chapter 3)

(2.2.2) $$\mathrm{ad}_a(x) = \Delta(a) \cdot x,$$

where $a, x \in A$, and $\Delta : A \to A \otimes A$ is the comultiplication. Similarly, we define the opposite adjoint action by

(2.2.3) $$\mathrm{ad}'_a(x) = \Delta'(a) \cdot x,$$

where $\Delta' = \sigma\Delta$ is the opposite comultiplication, and $\sigma : a \otimes b \mapsto b \otimes a$ is the permutation in $A \otimes A$.

Now suppose that $A = U_q\mathfrak{g}$, where $0 < q < 1$. We introduce the new generators

(2.2.4) $$E_i = X_i^+ K_i^{-1}, \quad F_i = X_i^- K_i.$$

Define the quantized simple reflections $\bar{s}_i$, $i = 1, 2, \ldots, n$, as in Definition 2.0.1.

PROPOSITION 2.2.1. *The following identities hold in $\mathbb{C}[G]_q^*$ (the Hopf algebra dual to $\mathbb{C}[G]_q$) provided i is not equal to j:*

(2.2.5) $$\bar{s}_i E_j \bar{s}_i^{-1} = c_{ij}^+ \left(\mathrm{ad}_{E_i}\right)^{-a_{ij}}(E_j),$$

(2.2.6) $$\bar{s}_i F_j \bar{s}_i^{-1} = c_{ij}^- \left(\mathrm{ad}'_{F_i}\right)^{-a_{ij}}(F_j),$$

(2.2.7) $$\bar{s}_i K_j \bar{s}_i^{-1} = K_j K_i^{-a_{ij}},$$

where $c_{ij}^\pm$ are some constants and $A = ((a_{ij}))_{i,j=1}^n$ the Cartan matrix of $\mathfrak{g}$.

PROOF. We will prove only (2.2.5), leaving the proof of the other formulas as an exercise to the reader. Consider the vector space

$$V_{ij} = \mathrm{Span}\{E_j, \mathrm{ad}_{E_i}(E_j), \ldots, (\mathrm{ad}_{E_i})^{-a_{ij}}(E_j)\}.$$

It is easy to see that V_{ij} is a $U_{q_i}\mathfrak{sl}(2,\mathbb{C})$-module with respect to the adjoint action, due to the quantum Serre relations. The Hopf subalgebra $U_{q_i}\mathfrak{sl}(2,\mathbb{C})$ in $U_q\mathfrak{g}$ is generated by $E_i, F_i, K_i^{\pm 1}$. Moreover, E_j is a lowest weight vector in V_{ij}, and $(\mathrm{ad}_{E_i})^{-a_{ij}}(E_j)$ is a highest weight vector.

By (2.1.4) and (2.1.12), we have a correctly defined action of $\bar{s}'_i$ in V_{ij}, so that we can write the following identity in $\mathbb{C}[G]^*_q$:

$$\mathrm{ad}_{\bar{s}'_i}(E_j) = c^+_{ij}\,(\mathrm{ad}_{E_i})^{-a_{ij}}(E_j),$$

where c^+_{ij} is a constant. By (2.2.2) and (2.1.13), we have

$$\mathrm{ad}_{\bar{s}'_i}(E_j) = \Delta(\bar{s}'_i) \cdot E_j = (R_i^{-1} \cdot (\bar{s}'_i E_j (\bar{s}'_i)^{-1}))(R_i \cdot 1).$$

Therefore,

$$R_i \cdot (\mathrm{ad}_{\bar{s}'_i}(E_j)) = (\bar{s}'_i E_j (\bar{s}'_i)^{-1})(R_i \cdot 1).$$

Now, the proposition follows from the following two identities in $\mathbb{C}[G]^*_q$:

$$R_i \cdot 1 = \sum_{n=0}^{\infty} \frac{(1 - q_i^{-2})^n}{(n)_{q_i^{-2}}!} E_i^n t_i^2 S(F_i^n),$$

$$E_i (\mathrm{ad}_{E_i})^{-a_{ij}}(E_j) = q^{(\alpha_i, \alpha_j)} (\mathrm{ad}_{E_i})^{-a_{ij}}(E_j) E_i.$$

The first identity follows from (2.1.4), (2.1.9), (2.1.11) (it is essentially a fact about $U_{q_i}\mathfrak{sl}(2,\mathbb{C})$). The second identity is a corollary of the quantum Serre relations (cf. Chapter 2, (1.2.3)), which imply that $(\mathrm{ad}_{E_i})^{1-a_{ij}}(E_j) = 0$. □

EXERCISE 2.2.2. Compute the constants $c^{\pm}_{ij}$ in (2.2.5) and (2.2.6).

The following proposition is an obvious corollary of (2.0.2).

PROPOSITION 2.2.3. Let $T_i(a) = \bar{s}_i a \bar{s}_i^{-1}$, for any $a \in U_q \mathfrak{g}$. Then we have that

$$T_i T_j T_i \cdots = T_j T_i T_j \cdots, \qquad i \neq j,$$

both sides consisting of m_{ij} factors, where m_{ij} is the same as in (2.0.2).

DEFINITION 2.2.4. The *quantum Weyl group* $\bar{W}_q$ is the Hopf subalgebra of $\mathbb{C}[G]^*_q$ generated by $U_q\mathfrak{g}$ and the quantized elements $\bar{w}$, $w \in W$.

Propositions 2.0.2 and 2.2.1 imply that $\bar{W}_q$ is generated by $E_i, F_i, K_i^{\pm 1}$, and $\bar{s}_i$ ($i = 1, 2, \ldots, n$). The correctness of the definition now follows from Proposition 2.1.4 which states that, indeed, $\bar{W}_q$ is a Hopf subalgebra in $\mathbb{C}[G]^*_q$.

REMARK 2.2.5. (1) The quantum Weyl group $\bar{W}_q$ is a quantum analog of the Hopf subalgebra of $\mathbb{C}[G]^*$ generated by $U\mathfrak{g}$ and $\bar{W}_1$.

(2) By Proposition 2.2.3, the operator T_w defined for any $w \in W$ by

$$T_w = T_{s_{i_1}} T_{s_{i_2}} \cdots T_{s_{i_k}},$$

does not depend on the choice of a reduced decomposition $w = s_{i_1} s_{i_2} \cdots s_{i_k}$.

2.3. Formal case. Using Chapter 3, Section 8, we can prove all the results of the previous sections of this chapter for algebras over the formal power series in h. For the formal quantizations we keep the same notation as the one used in the preceeding sections.

Namely, for any $w \in W$, we define a quantized element as a $\mathbb{C}[[h]]$-linear functional $\bar{w} : \mathbb{C}[G]_h \to \mathbb{C}[[h]]$ such that (2.0.1) holds (where π_{s_i} are regarded as irreducible $*$-representations of $\mathbb{C}[G]_h$).

EXERCISE 2.3.1. Prove the analogs of Propositions 2.0.2 and 2.1.1–2.1.3 in the formal case.

For any vertex i of the Dynkin diagram of $\mathfrak{g}$, we define

$$(2.3.1) \qquad R_i = \left(\sum_{n=0}^{\infty} \frac{(1-q_i^{-2})^n}{(n)_{q_i^{-2}}!} E_i^n \otimes F_i^n \right) q_i^{\frac{H_i \otimes H_i}{(\alpha_i, \alpha_i)}}$$

as an element of the h-adic completion of $(U_h\mathfrak{sl}(2,\mathbb{C})_i)^{\otimes 2} \subset (U_h\mathfrak{g})^{\otimes 2}$. Also, we define

$$(2.3.2) \qquad \bar{s}'_i = q^{-\frac{H_i^2}{2(\alpha_i, \alpha_i)}} \bar{s}_i.$$

EXERCISE 2.3.2. Prove Proposition 2.1.4 in the formal case, replacing t_i with $q^{-\frac{H_i^2}{2(\alpha_i, \alpha_i)}}$ in (2.1.15). Using (2.1.4), show that

$$(2.3.3) \qquad (\bar{s}'_i)^2 = e^{-\frac{hC_i}{2}} z_i,$$

where $e^{-\frac{hC_i}{2}}$ is the quantum Casimir element for $U_h\mathfrak{sl}(2,\mathbb{C})_i$ defined in Chapter 2, Corollary 4.1.2, and z_i is a group-like element ($\Delta(z_i) = z_i \otimes z_i$).

EXERCISE 2.3.3. Prove Propositions 2.2.1 and 2.2.3 in the formal case.

We define the *formal quantum Weyl group* $\bar{W}_h$ as the Hopf subalgebra in $\mathbb{C}[G]_h^*$ generated by $U_h\mathfrak{g}$ and the quantized elements $\bar{w}$ ($w \in W$). By definition, $\bar{W}_h$ is complete in the h-adic topology.

3. Quantum root vectors

3.1. Poincaré–Birkhoff–Witt theorem. Let $w_0 \in W$ be the longest element of the Weyl group, and $w_0 = s_{i_1} s_{i_2} \cdots s_{i_N}$ a reduced decomposition. By Proposition 2.0.2, we have that $\bar{w}_0 = \bar{s}_{i_1} \bar{s}_{i_2} \cdots \bar{s}_{i_N}$. Consider the following set of roots:

$$(3.1.1) \qquad \mathcal{D} = \{\alpha_{i_1}, s_{i_1} \alpha_{i_2}, \ldots, s_{i_1} s_{i_2} \cdots s_{i_{N-1}} \alpha_{i_N}\},$$

where $\alpha_1, \alpha_2, \ldots, \alpha_n$ are the simple roots of $\mathfrak{g}$.

EXERCISE 3.1.1. Prove that $\mathcal{D}$ coincides with the set Δ_+ of positive roots of $\mathfrak{g}$.

DEFINITION 3.1.2. A linear order L on the set Δ_+ of positive roots of $\mathfrak{g}$ is called *convex* if for any positive roots α and β such that $\alpha + \beta \in \Delta_+$ and $\alpha \prec_L \beta$, we have $\alpha \prec_L \alpha + \beta \prec_L \beta$.

EXERCISE 3.1.3. Show that the linear order on $\mathcal{D} = \Delta_+$ corresponding to the ordering of $\mathcal{D}$ in (3.1.1) from left to right, is a convex order. Moreover, prove that any convex order on Δ_+ comes from some reduced expression of w_0 in that way.

DEFINITION 3.1.4. Suppose that $\alpha = s_{i_1} s_{i_2} \cdots s_{i_{p-1}} \alpha_{i_p}$ ($p = 1, 2, \ldots, N$). We define the *root vectors* by

$$(3.1.2) \qquad E_\alpha = T_{i_1} T_{i_2} \cdots T_{i_{p-1}}(E_{i_p}), \qquad E_{\alpha_i} = E_i,$$
$$(3.1.3) \qquad F_\alpha = T_{i_1} T_{i_2} \cdots T_{i_{p-1}}(F_{i_p}), \qquad F_{\alpha_i} = F_i,$$

where $T_i : U_q\mathfrak{g} \to U_q\mathfrak{g}$ is the automorphism defined in Proposition 2.2.3.

EXERCISE 3.1.5. Show that $E_\alpha \in U_q\mathfrak{b}_+$ and $F_\alpha \in U_q\mathfrak{b}_-$.

REMARK 3.1.6. In the formal case, we define $E_\alpha, F_\alpha \in U_h\mathfrak{g}$ by the same formulas (3.1.2)–(3.1.3).

POINCARÉ–BIRKHOFF–WITT THEOREM. *Let $L : [1, N] \to \Delta_+$ be the convex linear order in Δ_+ described in Exercise 3.1.3. Then the following statements hold.*

(1) The monomials of the form $E_{L(1)}^{l_1} \cdots E_{L(N)}^{l_N} K_1^{m_1} \cdots K_n^{m_n}$ form a $\mathbb{C}$-linear basis in $U_q \mathfrak{b}_+$ when all l_i run over $\mathbb{Z}_+$ and all m_i run over $\mathbb{Z}$.

(2) The monomials of the form $E_{L(1)}^{l_1} \cdots E_{L(N)}^{l_N} K_1^{m_1} \cdots K_n^{m_n} F_{L(1)}^{p_1} \cdots F_{L(N)}^{p_N}$, $l_i, p_i \in \mathbb{Z}_+$, $m_i \in \mathbb{Z}$, form a $\mathbb{C}$-linear basis in $U_q \mathfrak{g}$.

(3) The monomials of the form $E_{L(1)}^{l_1} \cdots E_{L(N)}^{l_N} H_1^{m_1} \cdots H_n^{m_n}$ (resp. of the form $E_{L(1)}^{l_1} \cdots E_{L(N)}^{l_N} H_1^{m_1} \cdots H_n^{m_n} F_{L(1)}^{p_1} \cdots F_{L(N)}^{p_N}$) form a $\mathbb{C}[[h]]$-linear basis in the topological $\mathbb{C}[[h]]$-module $U_h \mathfrak{b}_+$ (resp. $U_h \mathfrak{g}$). Here $l_i, m_i, p_i \in \mathbb{Z}_+$.

PROOF. See [**Lus93**, Chapter 40]. □

REMARK 3.1.7. Strictly speaking, [**Lus93**, Chapter 40] contains the proofs for the version of $U_q \mathfrak{g}$ defined as a Hopf algebra over the field $\mathbb{Q}(q)$ of rational functions. It is not difficult to modify them for our purposes. We recommend it as an exercise to the reader to prove part (3) of the theorem using the deformation arguments and the Poincaré–Birkhoff–Witt theorem for $U \mathfrak{g}$.

3.2. Properties of quantum root vectors. From now on in this chapter, we will consider the case of the Hopf algebra $U_h \mathfrak{g}$ over $\mathbb{C}[[h]]$. The few exceptions we have will be clearly indicated below.

Let i be a vertex of the Dynkin diagram of $\mathfrak{g}$. We define

$$(3.2.1) \quad \hat{R}_i = \sum_{n=0}^{\infty} q^{-n^2(\alpha_i, \alpha_i)} \frac{(1 - q_i^{-2})^n}{(n)_{q_i^{-2}}!} E_i^n q^{-nH_i} \otimes F_i^n q^{nH_i}$$

as an element of the completion of $U_h \mathfrak{g} \otimes U_h \mathfrak{g}$ in the h-adic topology. By Exercise 2.3.2, the following identity holds for the elements of the formal quantum Weyl group $\bar{W}_h$:

$$(3.2.2) \quad \Delta(\bar{s}_i) = \hat{R}_i^{-1} (\bar{s}_i \otimes \bar{s}_i).$$

For any $\alpha \in \mathcal{D}$ such that $\alpha = s_{i_1} \cdots s_{i_{p-1}} \alpha_{i_p}$, we define the elements

$$(3.2.3) \quad \hat{R}_\alpha = T_{i_1} T_{i_2} \cdots T_{i_{p-1}}(R_{i_p}),$$

$$(3.2.4) \quad \hat{R}_{\prec \beta} = \prod_{\alpha \prec \beta} \hat{R}_\alpha,$$

where $T_i(a \otimes b) = T_i(a) \otimes T_i(b)$ for any $a, b \in U_h \mathfrak{g}$.

PROPOSITION 3.2.1. *Consider the canonical isomorphism $\mathfrak{h}^* \simeq \mathfrak{h}$ defined by the bilinear form $(\,,\,)$ on $\mathfrak{h}$. Let $H_\beta \in \mathfrak{h}$ be the image of a root $\beta \in \mathfrak{h}^*$ with respect to this isomorphism. Then the following identity holds:*

$$(3.2.5) \quad \Delta(E_\beta) = (\hat{R}_{\prec \beta})^{-1} (E_\beta \otimes 1 + q^{-H_\beta} \otimes E_\beta) \hat{R}_{\prec \beta}.$$

PROOF. The assertion follows immediately from (3.2.2)–(3.2.4). □

PROPOSITION 3.2.2. *Let $(U_h \mathfrak{b}_+)_{L(k)}$ be the h-adic complete $\mathbb{C}[[h]]$-submodule in $U_h \mathfrak{b}_+$ spanned by the monomials of the form $E_{L(1)}^{l_1} \cdots E_{L(K)}^{l_k}$, where $l_i \in \mathbb{Z}_+$. Then the following properties hold:*

(1) The element

$$\Delta(E_M^n) - (E_M \otimes 1 + q^{-H_M} \otimes E_M)^n$$

belongs to $(U_h \mathfrak{b}_+)_M \otimes U_h \mathfrak{b}_+$.

(2) $(U_h\mathfrak{b}_+)_M$ is a subalgebra in $U_h\mathfrak{b}_+$.

(3) We have
$$\Delta\left((U_h\mathfrak{b}_+)_M\right) \subset (U_h\mathfrak{b}_+)_M \otimes U_h\mathfrak{b}_+.$$

PROOF. The assertion follows from (3.2.5) and a more general result stated in the following theorem. □

THEOREM 3.2.3. (1) Let $\alpha \prec \beta$ be two positive roots of $\mathfrak{g}$. The we have that

(3.2.6) $$E_\alpha E_\beta - q^{-(\alpha,\beta)} E_\beta E_\alpha = \sum_{\alpha \prec \gamma_1 \prec \cdots \prec \gamma_j \prec \beta} c^+_{\bar{l},\bar{\gamma}} E^{l_1}_{\gamma_1} \cdots E^{l_j}_{\gamma_j},$$

(3.2.7) $$F_\alpha F_\beta - q^{-(\alpha,\beta)} F_\beta F_\alpha = \sum_{\alpha \prec \gamma_1 \prec \cdots \prec \gamma_j \prec \beta} c^-_{\bar{l},\bar{\gamma}} F^{l_1}_{\gamma_1} \cdots F^{l_j}_{\gamma_j},$$

where $c^{\pm}_{\bar{l},\bar{\gamma}} \in \mathbb{C}[[h]]$.

(2) The same identities hold in $U_q\mathfrak{g}$ with $c^{\pm}_{\bar{l},\bar{\gamma}} \in \mathbb{C}$.

PROOF. The proof given below works for both (1) and (2). In order to keep the notation consistent, we will consider the case of $U_h\mathfrak{g}$. The proof includes several steps.

Step 1. By the Poincaré–Birkhoff–Witt theorem, we have that

(3.2.8) $$E_\beta E_\alpha = \sum_{\bar{m} \in \mathbb{Z}^N_+} c_{\bar{m}} E^{m_1}_{L(1)} \cdots E^{m_N}_{L(N)}$$

for some $c_{\bar{m}} \in \mathbb{C}[[h]]$. We assume that $\alpha = L(r)$, $\beta = L(s)$, and $r < s$. We want to show that there are no nonzero summands on the right-hand side of (3.2.8) with $m_i > 0$ for $i > s$.

Step 2. The next step consists of a lemma.

LEMMA 3.2.4. (1) Let $w \in W$. If $w^{-1}\alpha_i > 0$, then $T^{-1}_w(E_i) \in U_h\mathfrak{b}_+$. And, if $w^{-1}\alpha_i < 0$, then $T^{-1}_w(E_i) \in U_h\mathfrak{b}_-$. Moreover, $T^{-1}_w(E_i)$ is a polynomial in E_α (resp. in F_α), where $\alpha \in \Delta_+$, and $T^{-1}_w(E_i) = P \cdot t$, where P is a polynomial in E_i (resp. in F_i), and $t \in U_h\mathfrak{h}$ is invertible.

(2) The analogous statement holds for $T^{-1}_w(F_i)$.

PROOF. A proof of the lemma can be found in [**Lus93**, Lemma 40.1.1]. Another way, suggested in [**LS90**], is based on the following identities, which can be proved similarly to (2.2.5) and (2.2.6):

(3.2.9) $$T^{-1}_i(E_j) = c^+_{ij} q_i^{-a_{ij}} (-\mathrm{ad}^{(q^{-1})}_{E_i})^{-a_{ij}}(E_j),$$

(3.2.10) $$T^{-1}_i(F_j) = c^-_{ij} q_i^{a_{ij}} (-\mathrm{ad}^{(q^{-1})'}_{F_i})^{-a_{ij}}(F_j).$$

Here we assume that $i \neq j$ and $\mathrm{ad}^{(q^{-1})}$ and $\mathrm{ad}^{(q^{-1})'}$ denote the quantum adjoint actions in the Hopf algebras $(U_h\mathfrak{g}, \Delta)$ and $(U_h\mathfrak{g}, \Delta')$ obtained by the automorphism $h \mapsto -h$, $E_i \mapsto E_i$, $F_i \mapsto F_i$, $H_i \mapsto H_i$. □

Step 3. Assume that there is a term $c_{\bar{m}} E^{m_1}_{L(1)} \cdots E^{m_p}_{L(p)}$ on the right-hand side of (3.2.8) with $p > s, m_p > 0, c_{\bar{m}} \neq 0$. We can assume that p is the largest number with this property.

Let $w_p = s_{i_1} \cdots s_{i_{p-1}}$. Apply $T^{-1}_{w_p}$ to both sides of (3.2.8). Since $w_p^{-1}(L(p))$ is a positive root and $w_p^{-1}(\alpha) \in \Delta_-, w_p^{-1}(\beta)$ are negative (for $p > s > r$), we get from Lemma 3.2.4 that $T_{w_p}^{-1}(E_\beta E_\alpha) \in U_h \mathfrak{b}_-$. Moreover, on the right-hand side, we obtain a sum in which all the terms but one belong to $U_h \mathfrak{b}_-$ (because of the choice of p). The single exceptional term must be of the form $\xi E_{L(p)}^{m_p}$, where $\xi \in U_h \mathfrak{b}_-$.

We can rewrite this term as $E_{L(p)}^{m_p} \xi + \eta$, where η does not contain $E_{L(p)}^{m_p}$ when represented as a linear combination of the monomials from the PBW basis, as described in Poincaré–Birkhoff–Witt theorem. Therefore, applying $T_{w_p}^{-1}$ to the right-hand side of (3.2.8) we cannot get an element from $U_h \mathfrak{b}_-$. This contradiction shows that for any $p > s$ we have $m_p = 0$ on the right-hand side of (3.2.8). Thus,

$$(3.2.11) \qquad E_\beta E_\alpha - a E_\alpha E_\beta = \sum_{\alpha \prec \gamma_1 \prec \cdots \prec \gamma_j \prec \beta} E_{\gamma_1}^{m_1} \cdots E_{\gamma_j}^{m_j},$$

where $a \in \mathbb{C}[[h]]$. Applying $T_{w_{r+1}}^{-1}$ to both sides of (3.2.11) (where $w_{r+1} = s_{i_1} \cdots s_{i_n}$), we obtain

$$(3.2.12)$$
$$T_{i_l} \cdots T_{i_{s-1}}(E_{i_s})(F_{i_n} q^{H_n}) - a F_{i_n} q^{H_n} T_{i_n} \cdots T_{i_s}(E_{i_s}) \in U_h \mathfrak{b}_-.$$

Rearranging the left-hand side of (3.2.12) in a linear combination of the elements of the PBW basis, we get that, in fact, $a = q^{-(\alpha,\beta)}$. □

4. The universal quantum R-matrix

We recall that in Section 2 of Chapter 2 the quantum double $\mathcal{D}(U_h \mathfrak{b}_+)$ was shown to be isomorphic to the (completed) tensor product $U_h \mathfrak{g} \otimes U_h \mathfrak{h}$. As a result, the universal quantum R-matrix for $U_h \mathfrak{g}$ is the image of the canonical element under the natural mapping $\mathcal{D}(U_h \mathfrak{b}_+) \otimes \mathcal{D}(U_h \mathfrak{b}_+) \to U_h \mathfrak{g} \otimes U_h \mathfrak{g}$ which is a Hopf algebra homomorphism. The purpose of this section is to describe this canonical element in terms of the PBW basis B constructed in Proposition 2.3.7. As a consequence, we give another proof of the above-mentioned description of $\mathcal{D}(U_h \mathfrak{b}_+)$.

4.1. The PBW-basis and the universal quantum R-matrix. Suppose that we have fixed a reduced decomposition of the longest element w_0 of the Weyl group. Let E_α, $\alpha \in \Delta_+$, be the root vector defined in Section 3. Introduce linear functionals ξ_i, $i = 1, 2, \ldots, n$ and η_α, $\alpha \in \Delta_+$, from $U_h \mathfrak{b}_+$ to $\mathbb{C}[[h]]$ given by

$$(4.1.1) \qquad \xi_i(H_i) = 0, \quad \xi_i(v) = 0 \text{ for any } v \in B \setminus H_i,$$
$$(4.1.2) \qquad \eta_\alpha(E_\alpha) = 1, \quad \eta_\alpha(v) = 0 \text{ for any } v \in B \setminus E_\alpha.$$

PROPOSITION 4.1.1. *The following formula holds:*

$$(4.1.3) \qquad \langle \eta_{\alpha(1)}^{k_1} \cdots \eta_{\alpha(N)}^{k_N} \xi_1^{p_1} \cdots \xi_n^{p_n}, E_{\alpha(1)}^{k_1'} \cdots E_{\alpha(N)}^{k_N'} H_1^{p_1'} \cdots H_n^{p_n'} \rangle$$
$$= \prod_{i=1}^n \delta_{p_i, p_i'} \prod_{j=1}^N \delta_{k_j, k_j'} \prod_{i=1}^n (p_i)! \prod_{j=1}^N (k_j)_{q_i^{-2}}!,$$

where $\langle , \rangle$ is the natural pairing between the dual spaces, and $(n)_t! = (1)_t \cdots (n)_t$ with $(n)_t = \frac{1-t^n}{1-t}$.

PROOF. The left-hand side of (4.1.3) can be rewritten as

$$(4.1.4) \qquad \langle \eta_{\alpha(1)} \otimes (\eta_{\alpha(1)}^{k_1-1} \cdots \xi_n^{p_n}), \Delta(E_{\alpha(1)})^{k_1'} \cdots \Delta(H_n)^{p_n'} \rangle.$$

It follows from (3.2.3) and Propositions 3.2.1 and 3.2.2 that (4.1.4) equals

$$\langle \eta_{\alpha(1)} \otimes (\eta_{\alpha(1)}^{k_1-1} \cdots \xi_n^{p_n}), (E_{\alpha(1)} \otimes 1 + q^{-H_{\alpha(1)}} \otimes E_{\alpha(1)})^{k_1'} \Delta(E_{\alpha(2)})^{k_2'} \cdots \Delta(H_n)^{p_n'} \rangle$$

$$= (k_1')_{q_1^{-2}} \langle \eta_{\alpha(1)}^{k_1-1} \cdots \xi_n^{p_n}, E_{\alpha(1)}^{k_1'-1} \Delta(E_{\alpha(2)})^{k_2'} \cdots \Delta(H_n)^{p_n'} \rangle$$

$$= \delta_{k_1, k_1'} (k_1')_{q_i^{-2}}! \langle \eta_{\alpha(1)}^{k_2} \cdots \xi_n^{p_n}, E_{\alpha(2)}^{k_2'} \cdots H_n^{p_n'} \rangle.$$

It is clear that one can continue and prove (4.1.3) by induction. □

COROLLARY 4.1.2. *The following identities hold in the Hopf algebra* $(U_h\mathfrak{b}_+)^*$ *which is dual to* $U_h\mathfrak{b}_+$, *where* $\eta_i = \eta_{\alpha_i}$ *correspond to the simple roots* α_i, $i = 1, 2, \ldots, n$:

- $[\xi_i, \eta_j] = -\delta_{ij} \frac{h}{2} \eta_j$;
- $[\xi_i, \xi_j] = 0$;
- $(\mathrm{ad}_{\eta_i})^{1-a_{ij}}(\eta_j) = 0$ *for any* $i \neq j$;
- $\Delta(\xi_i) = \xi_i \otimes 1 + 1 \otimes \xi_i$;
- $\Delta(\eta_i) = \eta_i \otimes 1 + e^{\sum_{k=1}^n (\alpha_i, \alpha_k) \xi_k} \otimes \eta_i$ *for any* $i, j = 1, 2, \ldots, n$.

PROOF. We leave it as an exercise to the reader to deduce the above indentities from Proposition 4.1.1 and Proposition 2.3.7 of Chapter 2. □

Let $(U_h\mathfrak{b}_+)^0$ be the QUE-dual Hopf algebra to $U_h\mathfrak{b}_+$ with the opposite comultiplication.

COROLLARY 4.1.3. *For any sequence* $\bar{c} = (c_1, \ldots, c_n)$ *of elements from* $\mathbb{C}[[h]]$ *such that* $c_i(0) \neq 0$ *for any* $i = 1, \ldots, n$, *there is an isomorphism* $\Phi_{\bar{c}}$ *between* $(U_h\mathfrak{b}_+)^0$ *and* $U_h\mathfrak{b}_-$ *given by*

$$\Phi_{\bar{c}}(\eta_j) = c_j F_j, \tag{4.1.5}$$

$$\Phi_{\bar{c}}\left(\exp\left(\sum_{k=1}^n (\alpha_i, \alpha_k)\xi_k\right)\right) = q^{H_i} \tag{4.1.6}$$

for any $i = 1, 2, \ldots, n$.

PROOF. The QUE-dual Hopf algebra $(U_h\mathfrak{b}_+)^0$ to $U_h\mathfrak{b}_+$ is defined in Section 2 of Chapter 2 as the h-adic completion of the subalgebra $\sum_{n=0}^\infty h^{-n}\mathfrak{m}^n$ of the QFSH-algebra $(U_h\mathfrak{b}_+)^*$. Here $\mathfrak{m}$ stands for the ideal topologically generated by ξ_i, η_i, $i = 1, 2, \ldots, n$. The corollary immediately follows from Corollary 4.1.2. □

REMARK 4.1.4. We recall that both Corollaries 4.1.2 and 4.1.3 have already been proved in Section 2 of Chapter 2. We recall them here just for completeness and consistency of the text.

EXERCISE 4.1.5. Let $\Phi_{\bar{c}}$ be as in Corollary 4.1.3 with $c_i = 1 - q_i^{-2}$, $i = 1, 2, \ldots, n$. Then $\Phi_{\bar{c}}(\eta_\alpha) = \left(1 - q_\alpha^{-2}\right) F_\alpha$ for any $\alpha \in \Delta_+$.

Hint: Define the automorphisms $\hat{T}_i$, $i = 1, 2, \ldots, n$ of $\mathcal{D}(U_h\mathfrak{b}_+)$ such that $\hat{T}_i$ coincides with T_i on $U_h\mathfrak{b}_+ \otimes 1$ and $\Phi_{\bar{c}}(\hat{T}_i x) = T_i \Phi_{\bar{c}}(x)$ for any $x \in \mathcal{D}(U_h\mathfrak{b}_+)$.

THEOREM 4.1.6. *The following formula holds for the universal quantum R-matrix of* $U_h\mathfrak{g}$:

$$R = \prod_{\alpha \in \Delta_+} \exp_{q_\alpha^{-2}}\left(\left(1 - q_\alpha^{-2}\right) E_\alpha \otimes F_\alpha\right) q^{t_0}, \tag{4.1.7}$$

where $t_0 \in \mathfrak{h} \otimes \mathfrak{h}$ is the canonical element corresponding to the invariant bilinear form on $\mathfrak{h}$, $q_\alpha = q^{\frac{(\alpha,\alpha)}{2}}$ and $\exp_t(x) = \sum_{n=0}^\infty \frac{x^n}{(n)_t!}$. The product in (4.1.7) is taken in the order described in Exercise 3.1.3.

PROOF. By Proposition 4.1.1, Corollary 4.1.3, and Exercise 4.1.5, we get that

$$
\begin{aligned}
R &= \sum_{m_i,k_i \in \mathbb{Z}_+} m_1! \cdots m_n! (k_1)_{q_{\alpha(1)}^{-2}}! \cdots (k_N)_{q_{\alpha(N)}^{-2}}! \\
&\quad \times E_{\alpha(1)}^{k_1} \cdots E_{\alpha(N)}^{k_N} H_1^{m_1} \cdots H_n^{m_n} \otimes \Phi\left(\eta_{\alpha(1)}^{k_1} \cdots \eta_{\alpha(N)}^{k_n} \xi_1^{m_1} \cdots \xi_n^{m_n}\right) \\
&= \prod_{\alpha \in \Delta_+} \exp_{q_\alpha^{-2}}\left((1 - q_\alpha^{-2}) E_\alpha \otimes F_\alpha\right) q^{t_0},
\end{aligned}
$$
(4.1.8)

and Theorem 4.1.6 is proved. □

COROLLARY 4.1.7. *Let $\bar{w}_0' = q^{-\frac{1}{2}\sum_{k=1}^n I_k^2} \bar{w}_0$, where $\{I_k\}$ is an orthonormal basis in $\mathfrak{h}$. Then*

$$\Delta(\bar{w}_0') = R^{-1}(\bar{w}_0' \otimes \bar{w}_0').$$
(4.1.9)

PROOF. By Exercise 2.3.2, formula (2.1.13) from Proposition 2.1.4 holds with R_i given by (2.3.1) (see also (3.2.2)). Then it is easy to see from (3.1.1)–(3.1.3) that $(\bar{w}_0' \otimes \bar{w}_0') \Delta(\bar{w}_0')^{-1}$ coincides with R given by (4.1.7). □

EXERCISE 4.1.8. Let $\bar{w}_0'$ be as in Corollary 4.1.7. Let

$$R_0 = (\bar{w}_0' \otimes \bar{w}_0') \Delta(\bar{w}_0')^{-1}.$$

Prove that $\Delta' = R_0 \Delta R_0^{-1}$ holds on the Chevalley generators E_i, F_i, H_i of $U_h\mathfrak{g}$ ($i = 1, 2, \ldots, n$). Use this fact to give another proof of (4.1.7).

Hint: There is a $\mathbb{C}[[h]]$-module M of infinite rank equipped with an action of $U_h\mathfrak{b}_+ \hat\otimes U_h\mathfrak{b}_-$ such that any $R \in U_h\mathfrak{b}_+ \hat\otimes U_h\mathfrak{b}_-$ satisfying the property $\Delta'(x) = R\Delta(x)R^{-1}$ for any $x \in U_h\mathfrak{g}$ is uniquely determined by its action on M.

4.2. The case of fixed q. Here we make some remarks on the case of fixed quantization parameter. Although the Hopf algebra $U_q\mathfrak{g}$ is not quasi-triangular, most of the computations done in the preceeding subsecton can be repeated in this case, too. The correct versions of the quantized elements of the quantum Weyl group belong to the Hopf algebra $\mathbb{C}[G]_q^*$ dual to $\mathbb{C}[G]_q$. We leave it to the reader to work out the details as well as the proof of the following proposition.

PROPOSITION 4.2.1. *Let*

$$\Theta = \prod_{\alpha \in \Delta_+} \exp_{q_\alpha^{-2}}\left((1 - q_\alpha^{-2}) E_\alpha \otimes F_\alpha\right)$$

be the part of the right-hand side of (4.1.7) without the Cartan factor q^{t_0}. Then Θ is a quasi R-matrix in the sense of Theorem 7.2.3 in Chapter 2.

5. Applications to the representations of the braid groups

EXERCISE 5.0.1. Let V be a finite-dimensional admissible $U_q\mathfrak{g}$-module. Then it carries a $\bar{W}_q$-module structure, where $\bar{W}_q$ is the quantum Weyl group (see Definition 2.2.4).

Let $V = L(\Lambda)$ be a finite-dimensional irreducible admissible $U_q\mathfrak{g}$-module with highest weight Λ, and $L(\Lambda) = \bigoplus_\lambda L(\Lambda)_\lambda$ its weight decomposition. Let us fix a weight λ and consider a vector space

$$E(\Lambda,\lambda) = \bigoplus_{w \in W} L(\Lambda)_{w\lambda}.$$

PROPOSITION 5.0.2. *The vector space $E(\Lambda,\lambda)$ carries a structure of B_n-module, where B_n is the braid group on n generators and n is the rank of $\mathfrak{g}$.*

PROOF. By Exercise 5.0.1 and (2.1.7), (2.2.7), we get $\bar{s}_i(E(\Lambda,\lambda)) \subset E(\Lambda,\lambda)$ for any quantized simple reflection $\bar{s}_i$, $i = 1, 2, \ldots, n$. Now the statement of the proposition follows from (2.0.2). □

Note that the spectrum of $\bar{s}_i$ as a linear automorphism of $E(\Lambda,\lambda)$ consists of finitely many points. One can try to compute it explicitly by using (2.1.4). The following proposition illustrates the results of such computations by a simple special case.

We recall that the Hecke algebra $H(t_1, \ldots, t_n; W)$ with $t_i \in \mathbb{C}^*$ is a $\mathbb{C}$-algebra with unity generated by $x_1, \ldots, x_n$ subject to the relations (2.0.2) with $\bar{s}_i$ replaced by x_i and

(5.0.1) $$(x_i - t_i)(x_i + 1) = 0.$$

PROPOSITION 5.0.3. *Let $V = L(\Lambda)$ be as above and $\Lambda(H_i) = \alpha_{\max}(H_i)$ ($i = 1, 2, \ldots, n$), where $\alpha_{\max}$ is the maximal root. Then $E(\Lambda,0)$ carries the structure of a module over the Hecke algebra $H\left(q_1^{-2}, \ldots, q_n^{-2}; W\right)$.*

PROOF. Consider $L(\Lambda)$ as a $U_{q_i}\mathfrak{sl}(2,\mathbb{C})$-module. It is well known that $L(\Lambda) = \bigoplus_k V_k^{(i)}$, where $V_k^{(i)}$ are all simple $U_{q_i}\mathfrak{sl}(2,\mathbb{C})$-modules with $\dim V_k^{(i)} \leq 3$. (This is true because $L(\Lambda)$ is a "quantum deformation" of the adjoint representation of $\mathfrak{g}$.) Then (2.1.4) tells us that the spectrum of the action of a semisimple element $\bar{s}_i$ restricted to $E(\Lambda,0)$ consists of two points: -1 and q_i^{-2}. This gives us relation (5.0.1) with $\bar{s}_i$ in place of x_i, and, therefore, a representation of $H\left(q_1^{-2}, \ldots, q_n^{-2}; W\right)$. □

REMARK 5.0.4. Proposition 5.0.3 is a "quantum analog" of the well-known result that the Weyl group W acts on the Cartan subalgebra $\mathfrak{h} \subset \mathfrak{g}$ which can be realized as $E(\Lambda,0) = L(\Lambda)_0$, the zero weight space for the adjoint representation of $\mathfrak{g}$.

6. Historical remarks

In this chapter we followed [**Soi91**, **SV88**, **LS91a**, **LS90**]. The relations between the Weyl group and the classical r-matrix used as motivation in Section 1 are due to Drinfeld. Formula (4.1.7) for the universal quantum R-matrix was first obtained in [**Soi91**] and later published with details in [**LS91a**, **LS90**]. Independently, it was obtained in [**KR90**]. Both [**Soi91**] and [**KR90**] were motivated by a similar result from [**Ros89**], where the case $\mathfrak{g} = \mathfrak{sl}(n,\mathbb{C})$ was considered. We mention [**KT91**], where another approach to (1.0.9) was proposed (which is closer to that of [**Ros89**]). Exercise 4.1.8 is based on [**KT92**].

We would like to mention also the book [**Lus93**] which presents a different approach to the automorphisms T_i, partially based on the earlier paper [**Lus88**]. Proposition 5.0.3 was first proved in [**Lus90**] in a different way.

We note that a definition of the quantum Weyl group in the case of infinite-dimensional Kac–Moody algebras has not been found yet. Nevertheless, the automorphisms playing the role of T_i can be defined in this case (cf. [**Lus93**]). Also, a formula similar to (1.0.9) can be proved, at least in some special cases (cf. [**KT92**]).

Quantum Weyl groups are discussed in several books besides [**Lus93**]. The discussions based on [**Soi91, SV88**] can be found, for example, in [**ChP95**] or [**Jos95**].

Bibliography

[BD84] A. A. Belavin and V. G. Drinfeld, *Triangle equations and simple Lie algebras*, Math. Phys. Reviews **4** (1984), 93–165.

[BGG73] I. N. Bernstein, I. M. Gelfand, and S. I. Gelfand, *Schubert cells and the cohomology of G/P*, Uspekhi Mat. Nauk **28** (1973), no. 3, 3–21; English transl. in Russian Math. Surveys **28** (1973).

[Bou89] N. Bourbaki, *Lie groups and Lie algebras*, Springer-Verlag, Berlin, New York, 1989.

[ChP95] V. Chari and A. Pressley, *Guide to quantum groups*, Cambridge Univ. Press, Cambridge, 1995

[Dix96] J. Dixmier, *Universal enveloping algebras*, Graduate Studies Math., vol. 11, Amer. Math. Soc., Providence, RI, 1996.

[Dri85] V. G. Drinfeld, *Hopf algebras and the quantum Yang-Baxter equation*, Dokl. Akad. Nauk SSSR **283** (1985), 1060–1064; English transl., Soviet Math. Dokl. **32** (1985), 264–267.

[Dri87] V. G. Drinfeld, *Quantum groups*, Proc. Internat. Congr. Math. (Berkeley, 1986), Amer. Math. Soc., Providence, RI, 1987, pp. 798–820.

[Dri89] V. G. Drinfeld, *On almost cocommutative Hopf algebras*, Algebra i Analiz **1** (1989), no. 2, 30–40; English transl., Leningrad Math. J. **1** (1990), 321–342.

[EK96] P. Etingof and D. Kazhdan, *Quantization of Lie bialgebras*, Selecta Math. **1** (1996), 1–41.

[FRT88] L. Faddeev, N. Reshetikhin, and L. Takhtajan, *Quantization of Lie groups and Lie algebras*, Algebraic Analysis, vol. 1, Academic Press, Boston, 1988, pp. 129–139.

[Fuk86] D. B. Fuks, *Cohomology of infinite-dimesional Lie algebras*, Consultants Bureau, New York, 1986.

[Gui80] A. Guichardet, *Cohomologie de groupes topologiques et des algèbres de Lie*, CEDIC, Paris, 1980.

[GS88] M. Gerstenhaber and S. Schack, *Algebraic cohomology and deformation theory*, Kluwer, Dordrecht, Boston, 1988.

[HL93] T. Hodges and T. Levasseur, *Primitive ideals of $\mathbf{C}_q[SL(3)]$*, Comm. Math. Phys. **156** (1993), 581–605.

[Jant96] J. Jantzen, *Lectures on quantum groups*, Amer. Math. Soc., Providence, RI, 1996.

[Jim86] M. Jimbo, *A q-difference analog of $U(g)$ and the Yang-Baxter equation*, Lett. Math. Phys. **11** (1986), 247–252.

[Jos95] A. Joseph, *Quantum groups and their primitive ideals*, Springer-Verlag, Berlin, New York, 1995.

[Kac90] V. G. Kac, *Infinite-dimensional Lie algebras*, 3rd edition, Cambridge University Press, Cambridge, New York, 1990.

[Kas95] C. Kassel, *Quantum groups*, Springer-Verlag, Berlin, New York, 1995.

[KR90] A. N. Kirillov and N. Yu. Reshetikhin, *q-Weyl group and a multiplicative formula for universal R-matrix*, Comm. Math. Phys. **134** (1990), 421–431.

[KT91] S. M. Khoroshkin and V. N. Tolstoy, *Universal R-matrix for quantized (super)algebras*, Comm. Math. Phys. **141** (1991), 599–617.

[KT92] S. M. Khoroshkin and V. N. Tolstoy *Universal R-matrix for quantized nontwisted affine Lie algebras*, Funktsional. Anal. i Prilozhen. **26** (1992), no. 3, 85–88; English transl., Functional Anal. Appl. **26** (1992), 69–71.

[Lan75] S. Lang, $SL_2(\mathbb{R})$, Addison-Wesley, Reading, MA, 1975.

[Lev91] S. L. Levendorskii, *Twisted function algebras on a compact quantum group and their representations*, Algebra i Analiz **3** (1991), no. 2, 180–198; English transl., St. Petersburg Math. J. **3** (1992), 405–423.

[LS90]   S. L. Levendorskii and Y. S. Soibelman, *Some applications of the quantum Weyl group*, J. Geom. Phys. **7** (1990), 141–170.

[LS91a] S. L. Levendorskii and Y. S. Soibelman. *The quantum Weyl group and a multiplicative formula for the R-matrix of a simple Lie algebra*, Funktsional. Anal. i Prilozhen. **25** (1991), 73–76; English transl., Functional Anal. Appl. **25** (1991), 143–145.

[LS91b] S. L. Levendorskii and Y. S. Soibelman, *Algebras of functions on compact quantum groups, Schubert cells, and quantum tori*, Comm. Math. Phys. **139** (1991), 141–170.

[Lu91]   J.-H. Lu, *Momentum mappings and reductions of Poisson actions*, Symplectic Geometry, Groupoids, and Integrable Systems (Berkeley, 1989), MSRI Publ., vol. 20, Springer-Verlag, New York, 1991, pp. 209–226.

[Lus88] G. Lusztig, *Quantum deformations of certain simple modules over enveloping algebras*, Adv. in Math. **70** (1988), 237–249.

[Lus90] G. Lusztig, *On quantum groups*, J. Algebra **131** (1990), 466–475.

[Lus93] G. Lusztig, *Introduction to quantum groups*, Birkhäuser, Boston, 1993.

[LW90]   J.-H. Lu and A. Weinstein, *Poisson Lie groups, dressing transformations, and Bruhat decompositions*, J. Diff. Geom. **31** (1990), 501–526.

[Mac71] S. MacLane, *Categories for the working mathematician*, Springer-Verlag, Berlin, New York, 1971.

[Maj95] S. Majid, *Foundations of quantum group theory*, Cambridge Univ. Press, Cambridge, 1995.

[Rie88] M. A. Rieffel, *Projective modules over higher-dimensional noncommutative tori*, Can. J. Math. **40** (1988), 257–338.

[Ros89] M. Rosso, *Analogue of PBW theorem and the universal R-matrix for $U_h sl(n+1)$*, Comm. Math. Phys. **124** (1989), 307–318.

[Rud73] W. Rudin, *Functional analysis*, McGraw-Hill, New York, 1973.

[STS85] M. A. Semenov-Tian-Shansky, *Dressing transformations and Poisson group actions*, Publ. Res. Inst. Math. Sci. **21** (1985), 1237–1260.

[Ser87] J.-P. Serre, *Complex semisimple Lie algebras*, Springer-Verlag, New York, 1987.

[SS93]   S. Shneider and S. Sternberg, *Quantum groups. From coalgebras to Drinfel'd algebras*, International Press, Cambridge, MA, 1993.

[Soi89] Y. S. Soibelman, *Algebra of functions on the quantum group $SU(n)$ and Schubert cells*, Dokl. Akad. Nauk SSSR **307** (1989), 41–45; English transl., Soviet Math. Dokl. **40** (1990), 34–38.

[Soi90] Y. S. Soibelman, *Algebra of functions on a compact quantum group and its representations*, Algebra i Analiz **2** (1990), no. 1, 190–212; English transl., Leningrad Math. J. **2** (1991), 193–225.

[Soi91] Y. S. Soibelman, *Quantum Weyl group and some of its applications* Proc. Winter School on Geometry and Physics (Srní, 1990), Rend. Circ. Mat. Palermo, Suppl. **26** (1991), no. 2, 155–181.

[Ste72] R. Steinberg, *Lectures on Chevalley groups*, Yale Univ. Press, New Haven, CT, 1972.

[SV86]   Y. S. Soibelman and L. L. Vaksman, *Algebra of functions on the quantum group $SU(2)$*, Preprint, Rostov-on-Don State Univ., Parts 1 and 2, 1986.

[SV88]   Y. S. Soibelman and L. L. Vaksman, *Algebra of functions on the quantum group $SU(2)$*, Funktsional. Anal. i Prilozhen. **22** (1988), no. 3, 1–14; English transl., Functional Anal. Appl. **22** (1988), 170–181.

[Wei83] A. Weinstein, *The local structure of Poisson manifolds*, J. Diff. Geom. **18** (1983), 523–557.

[Wo87a] S. Woronowicz, *Compact matrix pseudo-groups*, Comm. Math. Phys. **111** (1987), 613–665.

[Wo87b] S. Woronowicz, *Twisted $SU(2)$ group. An example of a noncommutative differential calculus*, Publ. Res. Inst. Math. Sci. **23** (1987), 117–181.